25th Jubilee Technical and Scientific Conference "Transport, Ecology – Sustainable Development (EKO Varna 2019)

IOP Conference Series: Materials Science and Engineering
Volume 614

Varna, Bulgaria
16 - 18 May 2019

Editors:

R. Hristov
R. Dimitrov

ISBN: 978-1-5108-9821-9
ISSN: 1757-8981

Printed from e-media with permission by:

Curran Associates, Inc.
57 Morehouse Lane
Red Hook, NY 12571

Some format issues inherent in the e-media version may also appear in this print version.

Institute of Physics
Dirac House, Temple Back
Bristol BS1 6BE UK

Phone: 44 1 17 929 7481
Fax: 44 1 17 920 0979

techtracking@iop.org

Additional copies of this publication are available from:

Curran Associates, Inc.
57 Morehouse Lane
Red Hook, NY 12571 USA
Phone: 845-758-0400
Fax: 845-758-2633
Email: curran@proceedings.com
Web: www.proceedings.com

Table of contents

Volume 614

Technical and Scientific Conference - Transport, Ecology - Sustainable Development

16–18 May 2019, Varna, Bulgaria

Accepted papers received: 13 August 2019
Published online: 24 September 2019

Preface

Papers

IOP Conf. Series: Materials Science and Engineering **614** (2019) 011001 doi:10.1088/1757-899X/614/1/011001

PREFACE

We are pleased to introduce the Proceedings of the 25th JUBILEE TECHNICAL AND SCIENTIFIC CONFERENCE "TRANSPORT, ECOLOGY – SUSTAINABLE DEVELOPMENT" EKO Varna 2019. The conference was held from May 16th to May 18th, 2019 in the city of Varna, Bulgaria. It was organised by Transport Engineering and Technologies Department at Technical University of Varna. The conference was attended by scientists and researchers from Bulgaria, Romania, Russia, Ukraine, Poland and Turkey. The opening was attended by representatives of Varna Regional Administration and the City Hall of Varna, as well as by the Rector of Technical University of Varna.

The main conference topics were:

- Vehicle design and technology
- Ecology of transport engines
- Bio fuels for internal combustion engines
- Road safety
- Electric vehicles and charging stations

The editors would like to thank all authors for their high-quality contributions and the members of the international programme committee for their commitment. We thank also to Academic Management of Technical University of Varna and to our commercial partners and exhibitors.

Official conference website: http://ecovarna.org

Editors
R. Hristov
R. Dimitrov

IOP Conf. Series: Materials Science and Engineering **614** (2019) 011002 doi:10.1088/1757-899X/614/1/011002

EKO VARNA

IOP Conf. Series: Materials Science and Engineering **614** (2019) 011003 doi:10.1088/1757-899X/614/1/011003

Peer review statement

All papers published in this volume of *IOP Conference Series: Materials Science and Engineering* have been peer reviewed through processes administered by the proceedings Editors. Reviews were conducted by expert referees to the professional and scientific standards expected of a proceedings journal published by IOP Publishing.

Mathematical simulation of stability of steady-state conditions for a road train

O Sakno[1*], D Moisia[2], T Kolesnikova[1], V Ollo[3]

[1]Department of Operation and Maintenance of Machines, Prydniprovs'ka State Academy of Civil Engineering and Architecture, 24A Chernyshevsky Street, Dnipro, 49600, Ukraine
[2]Faculty of Automotive and Mechanical Engineering, National Transport University, 1 Mykhaila Omelianovycha-Pavlenka Street, Kiev, 01010, Ukraine
[3]Department of Food and Supplies, Military Academy (Odesa), 10 Fontanska doroga, Odesa, 65009, Ukraine

*E-mail: sakno-olga@ukr.net

Abstract. The article is devoted to the application of the parameter continuation method for solving the problem of maneuverability of a three-trailer road train. The mathematical model of a three-trailer road train with a wheelless intermediate link is worked out. The analysis of the equilibrium conditions of the system is performed. The structural instability of the model of the road train in the neighbourhood of the linear motion is presented. The presence of stable circular steady-state conditions is proved when moving along circles of a sufficiently small radius. It is established that for a given longitudinal velocity there is a certain range of changes in the rotation angle of the steering wheels of the tractor unit, within which stable steady-state conditions may be. The analysis of the stability of the steady-state conditions at $\theta = 0.38$ rad. For the steady-state conditions, the configuration of the road train is drawn, where it can be seen that the without bearing link can only be oriented to the outer side of the turn. Also in the process of numerical integration, it is found that the resulting steady-state conditions are asymptotically stable.

1. Introduction

The mathematical model of the canonical road train is an object of research by many authors [1-3]. The results of the analysis are realized mainly for linear models [4-6]. The stability of circular steady-state conditions of motion of road train has not been studied sufficiently. Usually, a vehicular movement along the trajectory of major radius is considered [5, 7]. In this paper the software – computing base is developed. The base makes it possible to research the variety of the steady-state conditions of the road train model system control parameters such as the longitudinal velocity of freight traffic (the discrete parameter) and the wheel turning angle of the road train (the continuous parameter) are changed.

The paper is aimed at studying the nonlinear mathematical model of the three-trailer road train and shows possible restructurings of the configuration of the road train when the initial perturbation of the phase variable is changed. It requires the non-standard analysis.

It is well known that a wide class of problem which arises in several branches of pure and applied science can be studied in the general framework of the nonlinear equations. Due to their importance, several methods have been suggested and analyzed under certain conditions. It is tabular integration method, iteration method and parameter continuation method.

The object of the paper is a study the mobility and stability of the circular steady-state conditions of a model of a three-trailer road train and detection the set of steady-state conditions of the freight traffics by the parameter continuation method. The picture of the configuration of the road train is investigated when various traffic conditions.

We apply the parameter continuation method in order to study maneuverability of a three-trailer road train and periodic motions of stability, to find the bifurcation points. Technique allows finding out solutions step by step for each value of continuation parameter passing by the transitional process. It also opens the possibility to easily distinguish the instability zones and determine the bifurcation points. These methods show a good agreement of the results. In general, many authors have obtained bifurcation diagrams by time integration of the dynamics.

Tabular integration is an alternative to traditional integration by parts. The benefit of tabular integration is that it can save you a ton of time compared to integration by parts [8, 9]. This method requires that one of the functions be differentiable until it is zero. We have to integrate the other function every time differentiate the first function.

Iterative method uses successive approximations to obtain more accurate solutions to a linear system at each step. This method is the only choice for nonlinear equations [10-12].

2. The mathematical model of motion of a three-trailer road train

The mathematical model of motion of a road train is worked out. State variables are chosen for the most complete description and research of possible steady-state conditions of a three-trailer road train with without bearing intermediate member with rigid steering control (Figure 1) and without bearing semitrailer (Figure 2). Where *x*-*y* is fixed coordinate system; x_0-y_0 is moving coordinates, which is connected with the center of mass of a tractor unit; x_1-y_1 is moving coordinates, which is connected with the center of mass of a semi-trailer; x_2-y_2 is moving coordinates, which is connected with the center of mass of a trailer. The mathematical model helps to explain a system of motion of a road train and to study the effects of different components (variables), and to make predictions about behavior and stability of steady-state conditions.

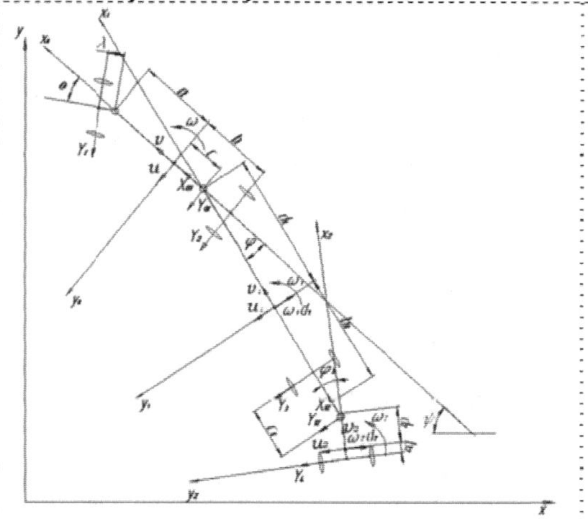

Figure 1. Basic design diagram of a three-trailer road train.

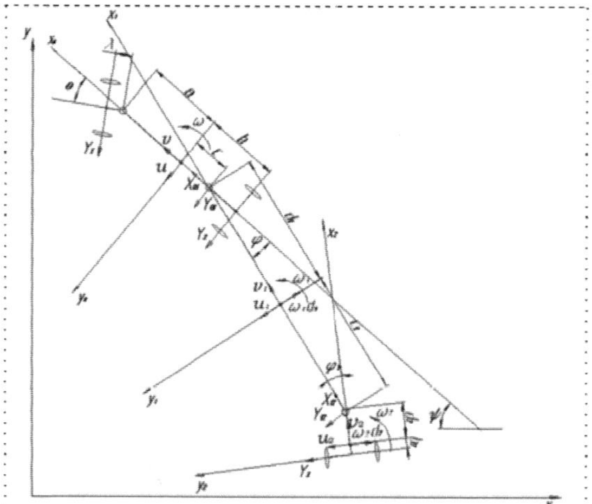

Figure 2. Basic design diagram of a three-trailer road train with without bearing semitrailer.

The front axle of a tractor unit turns by an angle θ. The connection between the links is carried out by cylindrical hinges, which allow free relative angulation of the links in the plane of their movement.

The position of each link is given by the coordinates x_i, y_i of its center of mass C_i and a course angle ψ_i (it is enclosed between the direct axis of the corresponding link and the x-axis of the fixed coordinate system).

Parameters of the system are the following:

v is longitudinal velocity component of the center of mass of the tractor unit

a is distance from the center of mass of the tractor unit to the points of attachment of the front axle of a tractor unit

b is distance from the center of mass of the tractor unit to the points of attachment of the rear axle of a tractor unit

c is distance from the center of mass of the tractor unit to the hitch point with a rear link

d_1 is distance from the center of mass of a semi-trailer to the hitch point

d_2 is distance from the center of mass of a trailer to the hitch point

$2K$ is overall width of the road train

k_f is friction coefficient

k_1, k_2 and k_3 are skid coefficient of axe (first and second and third axes)

χ_1, χ_2 and χ_3 are coefficient of friction to determining the force of lateral skid (first and second and third axes)

θ is wheel turning angle of subordinate module

Y_1, Y_2 and Y_3 are reduced lateral reaction of a roadbed on bearing axle skid (first and second and third axes).

The following values are set in the mathematical model:

C, C_1 and C_2 are centers of mass of the tractor unit, the semi-trailer and the trailer

m, m_1 and m_2 are masses of tractor unit, the semi-trailer and the trailer

I is central moment of inertia about vertical axis of the tractor unit

I_1 is central moment of inertia about vertical axis of the semi-trailer

I_2 is central moment of inertia about vertical axis of the trailer

φ is folding angle is formed between the longitudinal axes of the tractor unit and semi-trailer

φ_1 is folding angle is formed between the longitudinal axes of the tractor unit and trailer

$\omega = \dot{\psi}$, $\omega_1 = \dot{\psi}_1$ and $\omega_2 = \dot{\psi}_2$ are absolute angular velocities of the driving and driven links

Absolute angular velocities of links are the following:

$$\omega_1 = \omega - \dot{\varphi} \quad \text{and} \quad \omega_2 = \omega_1 - \dot{\varphi}_1$$

We set the absolute velocity of particles C, C_1 by resolutions in the unit vectors of the corresponding bases:

$$\begin{aligned}
&\mathrm{v_c} = \mathrm{i_0} \cdot \mathrm{v} + \mathrm{j_0} \cdot \mathrm{u} \\
&\mathrm{v} = \dot{\mathrm{x}} \cdot \cos\psi + \dot{y} \cdot \sin\psi \\
&\mathrm{v_1} = \mathrm{v} \cdot \cos\varphi - (u - \omega \cdot c) \cdot \sin\varphi \\
&\mathrm{v_2} = \mathrm{v_1} \cdot \cos\varphi_1 - (u_1 - \omega_1 \cdot b_1) \cdot \sin\varphi_1 \\
&\mathrm{v_{c1}} = \mathrm{i_1} \cdot \mathrm{v_1} + \mathrm{j_1} \cdot \mathrm{u_1} \\
&u = -\dot{x} \cdot \sin\psi + y \cdot \cos\psi \\
&u_1 = \mathrm{v} \cdot \sin\varphi + (u - \omega \cdot c) \cdot \cos\varphi - \omega_1 \cdot d_1 \\
&u_2 = \mathrm{v_1} \cdot \sin\varphi_1 + (u_1 - \omega_1 \cdot b_1) \cdot \cos\varphi_1 - \omega_2 \cdot d_2
\end{aligned} \tag{1}$$

The differential equation system of motion of a road train describes the change in the phase variables (u, ω, φ, φ_1, $\triangleleft$, $\triangleleft_1$). Where u is lateral velocity of center of mass of the tractor unit (quasi-velocity); U is derivative of u in the moving coordinates; Ω is angular acceleration relative to the vertical axis; $\triangleleft$ – velocity of folding angle φ; $\triangleleft_1$ – velocity of folding angle φ_1; PP – angular acceleration of a semi-trailer relative to intrinsic vertical axis; PP_1 – angular acceleration of a trailer relative to intrinsic vertical axis.

There are a lot of theories of rolling of elastically deformed wheels. The axiomatic is the most widespread, which asserts that a lateral reaction Y_i of a roadbed is applied at the tooth bearing center of the rolling elastic wheel. It is a function of wheel slip angle δ_i.

The reduced angles of lateral skid of wheel axles are determined by expressions:

$$\delta_1 = \theta - arctg\left[\frac{u + a\cdot\omega}{v}\right]$$
$$\delta_2 = arctg\left[\frac{-u + b\cdot\omega}{v}\right] \quad (2)$$
$$\delta_3 = arctg\left[\frac{-u_1 + b_1\cdot\omega_1}{v_1}\right]$$

Dependences of forces of lateral skid possess an empirical origin [13]. It is determined by expressions:

$$Y_i = \frac{k_i\cdot\delta_i}{\sqrt{[1+(k_i\cdot\delta_i/\chi_i\cdot Z_i]^2}} \quad (3)$$

Where Z_i is reaction of footprint on axes.

Values Z_1, Z_2, Z_3 are determined by expressions:

$$Z_1 = \frac{1}{l}\cdot\left[m\cdot g\cdot b - (m_2 + m_1)\cdot g\cdot\frac{b_1}{L_1}\cdot(c-b)\right]$$
$$Z_2 = \frac{1}{l}\cdot\left[m\cdot g\cdot a + (m_2 + m_1)\cdot g\cdot\frac{b_1}{L_1}\cdot(c+a)\right] \quad (4)$$
$$Z_3 = (m_2 + m_1)\cdot g\cdot\frac{d_1}{L_1};\ l = a + b;\ L_1 = d_1 + b_1 + d_2 + b_2$$

The motion equations of tractor unit are formed as:

$$m\cdot(V - u\cdot\omega) - X_{01} + Y_1\cdot\sin(\theta) = 0$$
$$m\cdot(U + v\cdot\omega) - Y_{01} - Y_1\cdot\cos(\theta) - Y_2 = 0 \quad (5)$$
$$I\cdot\Omega + Y_{01}\cdot c - Y_1\cdot a\cdot\cos(\theta) + Y_2\cdot b = 0$$

The motion equations of semi-trailer are formed as:

$$\mathrm{m}_1\cdot(\mathrm{V}_1 - \mathrm{u}_1\cdot\omega_1) + X_{01}\cdot\cos(\varphi) - Y_{01}\cdot\sin(\varphi) - X_{12} = 0$$
$$m_1\cdot(U_1 + v_1\cdot\omega_1) + \mathrm{X}_{01}\cdot\sin(\varphi) + Y_{01}\cos(\varphi) - Y_{12} - Y_3 = 0 \quad (6)$$
$$I_1\cdot\Omega_1 + X_{01}\cdot d_1\cdot\sin(\varphi) + Y_{01}\cdot d_1\cdot\cos(\varphi) + Y_{12}\cdot b_1 + Y_3\cdot(b_1 - c_1) = 0$$

The motion equations of without bearing semitrailer are formed as:

$$\mathrm{m}_1(\mathrm{V}_1 - \mathrm{u}_1\omega_1) + X_{01}\cos(\varphi) - Y_{01}\sin(\varphi) - X_{12} = 0,$$
$$m_1(U_1 + v_1\omega_1) + \mathrm{X}_{01}\sin(\varphi) + Y_{01}\cos(\varphi) - Y_{12} = 0, \quad (7)$$
$$J_1\Omega_1 + X_{01}\,d_1\sin(\varphi) + Y_{01}d_1\cos(\varphi) + Y_{12}b_1 = 0.$$

The motion equations of trailer are formed as:

$$\begin{aligned}
&\mathrm{m}_2\cdot(\mathrm{V}_2-\mathrm{u}_2\cdot\omega_2)+X_{12}\cdot\cos(\varphi_1)-Y_{12}\cdot\sin(\varphi_1)=0\\
&m_1\cdot(U_1+v_1\cdot\omega_1)+\mathrm{X}_{12}\cdot\sin(\varphi_1)+Y_{12}\cdot\cos(\varphi_1)-Y_4=0\\
&I_2\cdot\Omega_2+X_{12}\cdot d_2\cdot\sin(\varphi_1)+Y_{12}\cdot d_2\cdot\cos(\varphi_1)+Y_4\cdot b_2=0
\end{aligned}\tag{8}$$

The motion equations of without bearing semitrailer are formed as:

$$\begin{aligned}
&\mathrm{m}_2(\mathrm{V}_2-\mathrm{u}_2\omega_2)+X_{12}\cos(\varphi_1)-Y_{12}\sin(\varphi)=0,\\
&m_1(U_1+v_1\omega_1)+\mathrm{X}_{12}\sin(\varphi_1)+Y_{12}\cos(\varphi_1)-Y_3=0,\\
&J_1\Omega_1+X_{12}d_2\sin(\varphi_1)+Y_{12}d_2\cos(\varphi_1)+Y_3b_2=0.
\end{aligned}\tag{9}$$

Internal forces Y_{01}, X_{01}, Y_{12}, X_{12} are eliminated from Equations (5), (6), (7), (8) and (9). The system of nonlinear differential equation is got more generally by variable values v, u, ω, φ, φ_1:

$$\mathrm{f}(\mathrm{v},\mathrm{u},\omega,\varphi,\varphi_1,\mathrm{V},\mathrm{U},\Omega,\Phi,\Phi_1,\mathrm{PP},\mathrm{PP}_1,\theta)=0\tag{10}$$

With steady-state condition of motion, the values of all phase variables become constant. Thus, the set of steady-state conditions is assessed by system (10), where $v=\mathrm{const}$ and $\mathrm{V}=\mathrm{U}=\Omega=\Phi=\Phi_1=\mathrm{PP}=\mathrm{PP}_1=0$.

The system is got more general view:

by variable value

$$\begin{aligned}
u &\quad \mathrm{f}(\mathrm{u},\omega,\varphi,\varphi_1,\theta)=0\\
\omega &\quad \mathrm{f}(\mathrm{u},\omega,\varphi,\varphi_1,\theta)=0\\
\varphi &\quad \mathrm{f}(\mathrm{u},\omega,\varphi,\varphi_1,\theta)=0\\
\varphi_1 &\quad \mathrm{f}(\mathrm{u},\omega,\varphi,\varphi_1,\theta)=0
\end{aligned}\tag{11}$$

3. Analysis of stability of a three-trailer road train with bearing semitrailer

The system is analyzed for the presence of circular steady-state conditions and to assess of the road train.

An example is reduced for curve construction of real solutions of the quadratic equation when one parameter (x) changes, using the parameter continuation method. The equation is formed as:

$$x^2+y^2-1=0\tag{12}$$

Solution real set of the equation is presented in the form of implicit function $G(x,y) = 0$. In plane (x,y) it is presented some curve, which passes across origin of coordinates (obvious solution of equation $\{x = 0, y = 0\}$). The curve is parameterized by natural parameter. It is arc length

$$s\left(ds=\sqrt{dx^2+dy^2}\right).$$

The system of differential equations is obtained by the parameter continuation method (x) [14]:

$$\frac{dx}{ds}=\frac{\dfrac{\partial G}{\partial y}}{\sqrt{\left(\dfrac{\partial G}{\partial y}\right)^2+\left(\dfrac{\partial G}{\partial x}\right)^2}}$$

$$\frac{dy}{ds}=-\frac{\frac{\partial G}{\partial x}}{\sqrt{\left(\frac{\partial G}{\partial y}\right)^2+\left(\frac{\partial G}{\partial x}\right)^2}}$$

Cubic parabola of Equation (12) is drawn using the parameter continuation method (Figure 3).

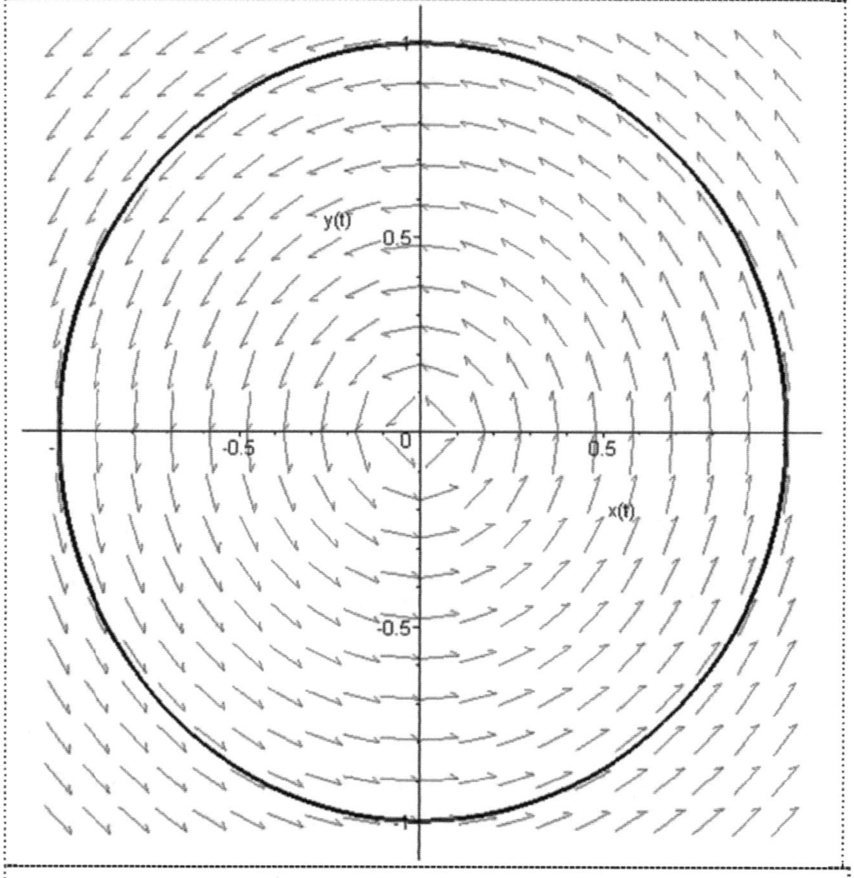

Figure 3. Cubic parabola (horizontal axis is the parameter x of the Equation (12); vertical axis is the parameter y of the Equation (12)).

4. Results of numerical analysis of the mathematical model of the road train with without bearing semitrailer

The system is examined for the presence of circular steady-state conditions and the configuration of the road train is assessed.

The system of Equations (11) is used to determine the steady-state conditions by the iterative method. System of Equations (10) is used for numerical integration. It is necessary to prove the instability of this model in the neighborhood of the linear motion. Jacobi matrix of Equations (11) is equated by phase variables:

$$\left[\frac{k_1+k_2+k_3}{v},\frac{m_1v^2+m_2v^2+k_1a-k_3c-k_3d_1-k_3c_1-k_3d_2-k_3b_2-k_2b+mv^2}{v},k_3,k_3\right]$$

$$\left[\frac{k_3c-k_1a+k_2b}{v},-\frac{-k_3cd_1-k_2b^2-k_3c^2-k_1a^2-ck_3c_1+cm_1v^2+cm_2v^2-k_3cd_2-k_3b_2c}{v},-k_3c,-k_3c\right]$$

$$\left[-\frac{k_3\cdot(d_1+c_1)}{v},-\frac{-ck_3d_1-ck_3c_1-k_3d_1^2-k_3c_1^2+d_1m_1v^2+d_1m_2v^2+c_1m_2v^2-2c_1k_3d_1-d_2k_3d_1-d_2k_3c_1-b_2k_3d_1-b_2k_3c_1}{v},\right.$$
$$\left.-k_3d_1-k_3c_1,k_3\cdot(-d_1-c_1)\right]$$

$$\left[-\frac{k_3\cdot(d_2+b_2)}{v},-\frac{-d_2m_2v^2-d_2^2k_3-b_2^2k_3-d_2k_3c-d_2k_3d_1-d_2k_3c_1-b_2k_3c-d_1k_3b_2-b_2k_3c_1-2d_2k_3b_2}{v},\right.$$
$$\left.-k_3d_2-k_3b_2,-k_3d_2-k_3b_2\right]$$

The determinant of the matrix is equal to the free term of the characteristic equation. A necessary condition for stability according to A. Stodol is that the free term of the characteristic equation must be greater than zero. The value of the longitudinal velocity v is critical, which determines the determinant of the matrix to zero (when this velocity is reached, a loss of stability occurs). Thus, if determinant is calculated the analytically, equated it to zero, and expressed v from the resulting equality, we obtain the expression of the critical velocity in the analytic form. Calculations are shown, that the determinant of the matrix is identically zero, and therefore it does not depend on v. So the model is structurally unstable in the neighbourhood of the rectilinear mode, and the value of the critical velocity is zero. Attempts to determine stable steady-state conditions when moving along circles of a sufficiently large radius also revealed structural instability.

Circular steady-state conditions are determined when the road train moves along circles of a sufficiently small radius. For the received steady-state conditions, the characteristic configuration of the road train is used.

The following parameters are defined:

$\chi_i = 0.8$, $K = 1.5m$, $\mathrm{k}_1 = 300000$ H, $\mathrm{k}_2 = 800000$ H, $\mathrm{k}_3 = 1000000$ H,

m = 6250 kg, $\mathrm{m}_2 = 6160$ kg, a = 1.27 m, b = 2.33m, c = 1.6 m, $\mathrm{b}_1 = 2.8$ m, $\mathrm{d}_1 = 4$ m, v = 5 m/s, $\mathrm{c}_1 = 3.79$ m;

$\theta = 0.38\ rad$; $J = 10000\ kg\cdot m^2$, $J_1 = 25000\ kg\cdot m^2$, $J_2 = 200\ kg\cdot m^2$

These parameters are corresponded to the steady circular steady-state conditions:

$\{\mathrm{u}^* = 1.041\mathrm{m/s},\ \omega^* = 0.511 rad/s,\ \varphi^* = -7.898 rad, \varphi_1{}^* = -4.676 rad\}$

The stability of the steady-state conditions is confirmed the eigenvalue spectrum:

eigv := -0.005288260944 + 0.6250834805 *I*, -14.17459637 + 1.386043145 *I*,

-33.64463075, -123.1874797, -14.17459637 − 1.386043145 *I*,

-0.005288260944 − 0.6250834805 *I*

The roots of the performance equation of the system in variations have negative real parts. According to the Lyapunov theorem, the circular traffic condition is asymptotically stable. Besides, the configuration of the road train has the essential singularity. This wheelless link is oriented to the outer side of the center of mass of the tractor unit (Figure 4). The ordinate of point *C* is determined the radius of the circle on which the center of mass of the tractor unit is moved (Y_c = 9.99 m), the rotation center is the coordinate origin. In this case, the velocity vector of the center of mass of the tractor unit is parallel to the abscissa axis, so radius of rotation of the tractor unit is perpendicular to the same axis. This result is expected, since, with a different configuration of the road train, it is not possible to balance the centrifugal force of the without bearing semitrailer. The stability of these conditions is asymptotic. This is confirmed by the results obtained by the numerical integration of the differential equations of the road train movement (Figure 5).

The Figure 5a is illustrated the process of entering into steady-state conditions of the road train, whereupon the radius of curvature of the trajectory remains almost constant $R \approx 10$ m, but there is the displacement in the rotation center. This fact is confirmed by the geometric construction of the

IOP Conf. Series: Materials Science and Engineering **614** (2019) 012001 doi:10.1088/1757-899X/614/1/012001

trajectory in the steady-state conditions by the iterative method and the radius value in this case is R = 9.986 m.

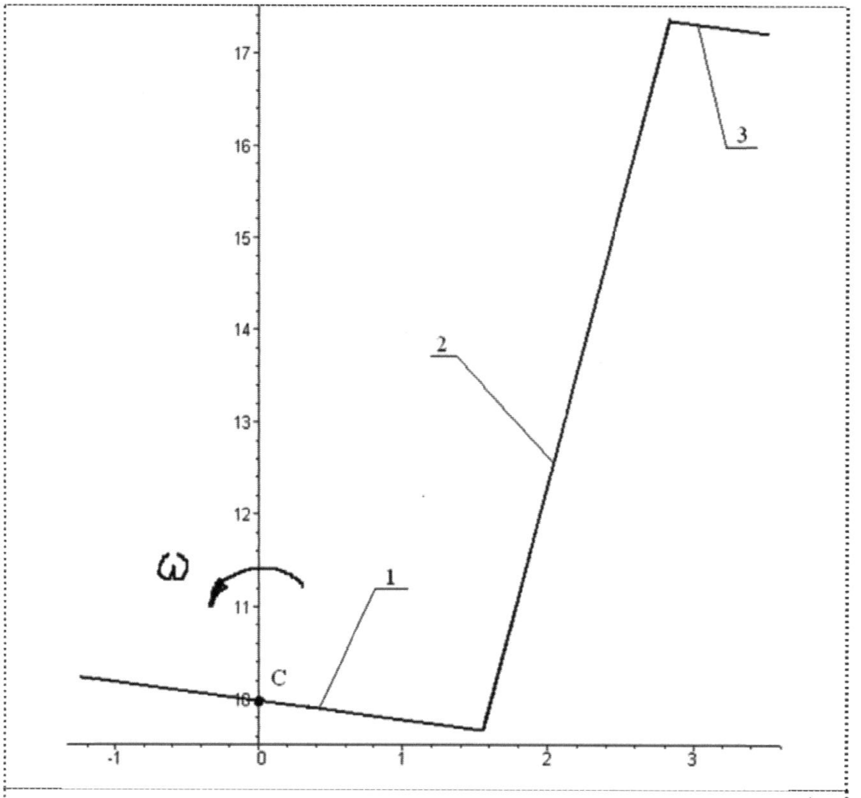

Figure 4. Configuration of road train in the road plane [m] when steady-state condition: 1 – longitudinal axis of the tractor; 2 – longitudinal axis of the semi-trailer, 3 – longitudinal axis of the bearing bogie; *C* – center of mass of the tractor unit.

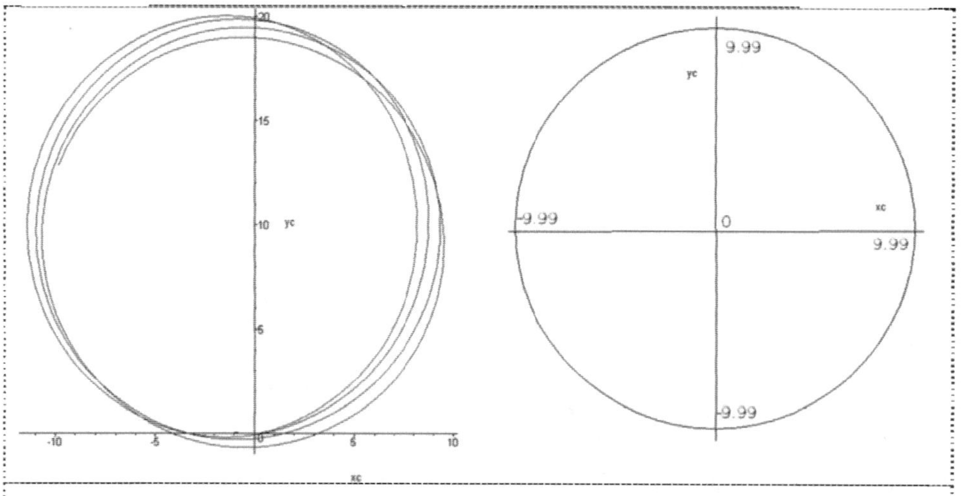

a) b)

Figure 5. The trajectory of the center of mass of the tractor unit in the road plane [m]: a) by math software Maple; b) by geometrically.

IOP Conf. Series: Materials Science and Engineering **614** (2019) 012001 doi:10.1088/1757-899X/614/1/012001

Graphs of phase dependence of phase variable on time are presented in the Figure 6. The figure fully is confirmed the asymptotic stability in the steady-state conditions. The Figures 6a, 6b and 6c are shown phase trajectories that have a vibrational damped character. The Figure 6d is shown that the folding angle of the without bearing link is less than zero. This is corroborated the orientation of the semitrailer to the outside of the turn.

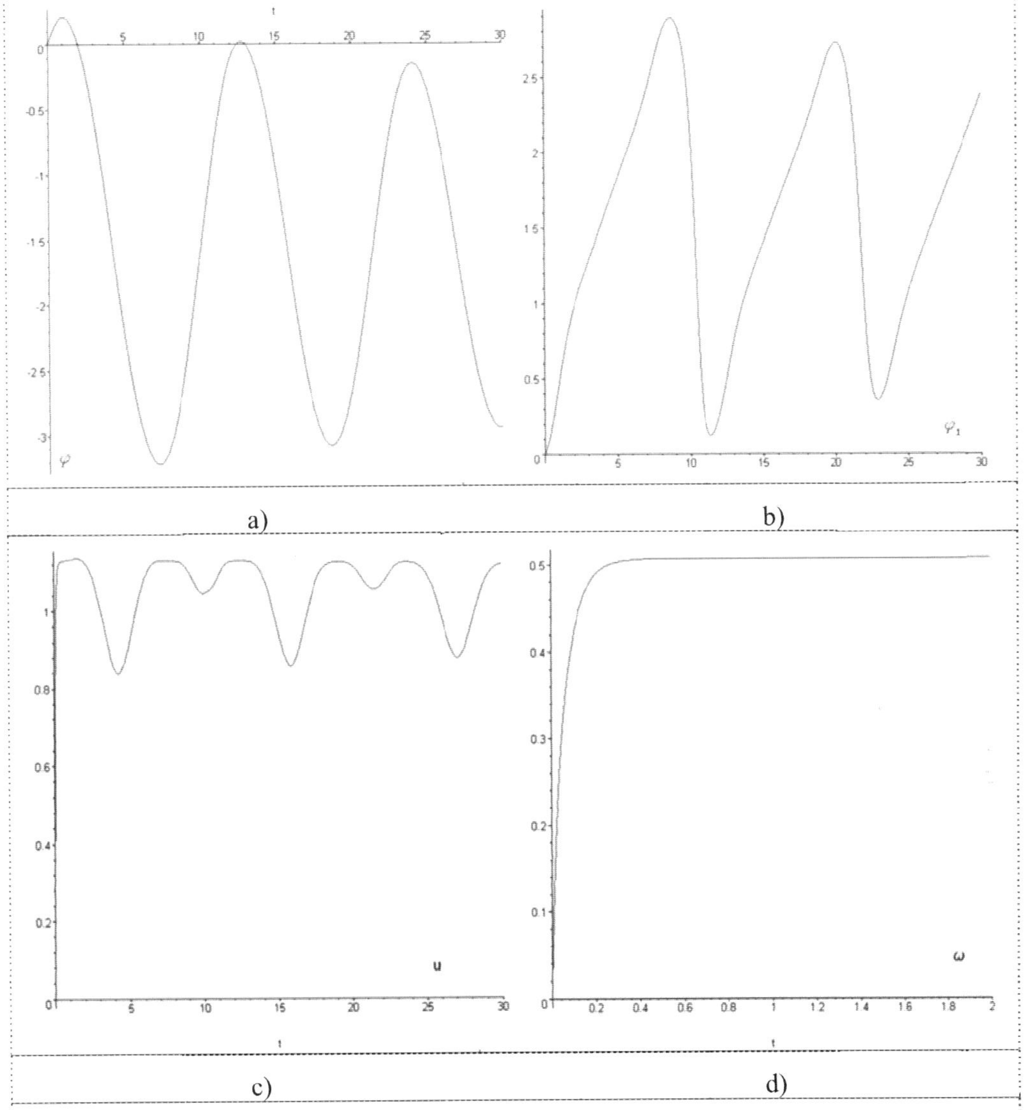

Figure 6. Phase trajectory: a) folging angle of the semitrailer; b) folging angle of the s bearing bogie; c) transverse velocity; d) angular velocity.

The change of the configuration of the road train at different values of the steering wheel turning angle of the tractor unit is analysed. The road train configurations are presented in Figure 7. The construction is based on the values of the steady-state conditions by the iterative method.

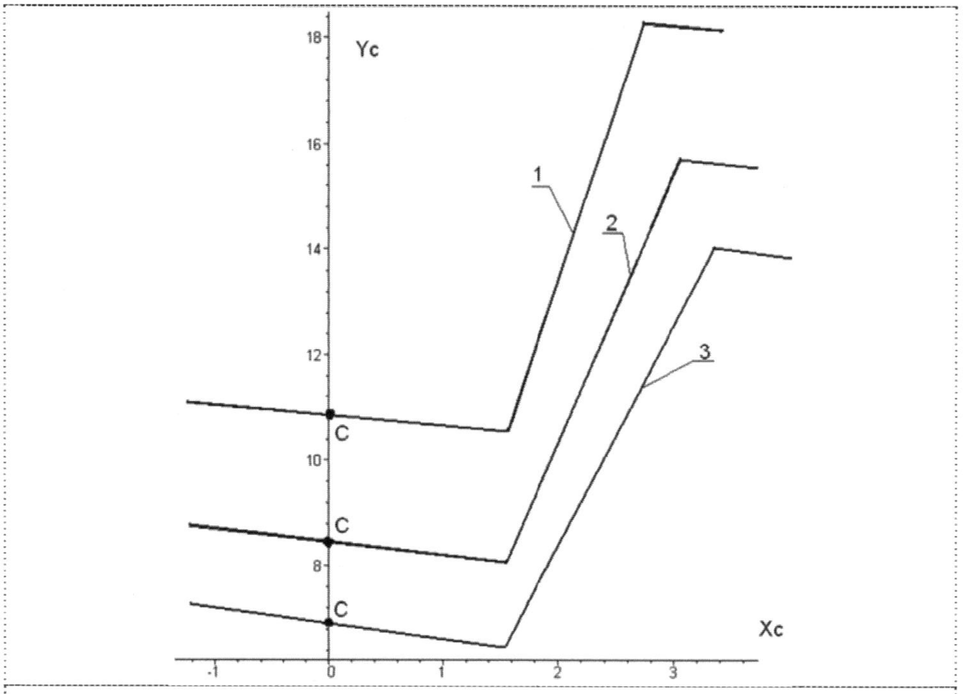

Figure 7. Configuration of the road train at different values of the wheel turning angle of the front link: 1 – θ = 0.35 rad; 2 – θ = 0.45 rad; 3 – θ = 0.55 rad.

The Figure 7 is showed that the configurations are not significantly different, the radius of rotation of the road train is changed (it is the ordinates of points *C*) and the folding angles is changed slightly.

When analyzing the mathematical model, stable steady-state conditions are realized in the range of 0.1 rad < θ < 0.9 rad.

5. Conclusions

The mathematical model of a three-trailer road train with a wheelless intermediate link is worked out. The analysis of the equilibrium conditions of the system is performed. The structural instability of the model of the road train in the neighbourhood of the linear motion is presented. The presence of stable circular steady-state conditions is proved when moving along circles of a sufficiently small radius. It is established that for a given longitudinal velocity there is a certain range of changes in the rotation angle of the steering wheels of the tractor unit, within which stable steady-state conditions may be. The analysis of the stability of the steady-state conditions at θ = 0.38 rad. For the steady-state conditions, the configuration of the road train is drawn, where it can be seen that the without bearing link can only be oriented to the outer side of the turn. Also in the process of numerical integration, it is found that the resulting steady-state conditions are asymptotically stable.

Acknowledgments

The authors would like to express their gratitude to Prof. Volodymyr Verbitskii and Prof. Volodymyr Sakhno.

References

[1] Ritzen P, Roebroek E, Wouw N, Jiang Z-P and Nijmeijer H 2016 *IEEE Transactions on Control Systems Technology* **24 (4)** 1240-1252.

[2] Verbitskii V and Lobas L 1987 *Applied Mathematics and Mechanics* **3** 418-425 (in Russian).

[3] Kayacan E, Kayacan E, Ramon H and Saeys W 2014 *Mechatronics* **24** **(8)** 926-933.

[4] Lobas L and Verbitskii V 1990 *Qualitative and analytical methods in the dynamics of wheeled vehicles* (Kiev: Naukova Dumka) p 232 (in Russian).

[5] Ellis J 1969 *Vehicle Dynamics* (London: Business Books) p 243.

[6] Shinohara Y 1972 *Publications of the Research Institute for Mathematical Sciences of Kyoto University* **8** **(1)** 13-42.

[7] Bazhenov V A, Pogorelova O S and Postnikova T G 2014 *Opir materialiv i teoria sporud-strength of materials and theory of structures* **93** 110-117.

[8] Azevedo R C, Peres J and von Stosch M 2015 *Engineering Applications of Artificial Intelligence* **38** 24-33.

[9] Bardella L, Carini A and Genna F 2003 *Computers & Structures* **81** **(24–25)** 2361-2372

[10] Noor M A, Noor K I, Al-Said E and Waseem M 2010 *Mathematical Problems in Engineering* 12 p

[11] Ujević N 2007 *J. of Computational and Applied Mathematics* **201** **(1)** 208-216.

[12] Mohyud-Din S T, Sikander W, Khan U and Ahmed N 2017 *J. of the Association of Arab Universities for Basic and Applied Sciences* **24** 191-197.

[13] Rocard Y. 1954 *L'instabilité en mécanique, automobiles-avions-ponts suspendus* (Paris: Masson et Cie).

[14] Sakno O, Moisia D, Kolesnikova T, Ollo V and Antropov A 2019 *Transport, ecology – sustainable development* Varna: Technical University Of Varna, Department Of Transport Technique And Technology.

IOP Conf. Series: Materials Science and Engineering **614** (2019) 012002 doi:10.1088/1757-899X/614/1/012002

Assortment analysis of shutter valves for spark ignition engines timing systems

J Matla

Wrocław University of Science and Technology, Faculty of Mechanical Engineering Division of Vehicles Engineering, ul. Braci Gierymskich 164, 51-640 Wrocław, POLAND

jedrzej.matla@pwr.edu.pl

Abstract. Continually tightened emission restrictions together with limited crude oil resources are factors driving more efficient fuel management strategies implementation. Such operations results in increase of internal combustion engine efficiency due to change in geometry and materials of existing systems as well as introducing new solutions. Over the years many approaches were taken. One of them was timing valve geometry optimization, which in case of conventional poppet valves is burden with number of restrictions. At this place some nonconventional designs might be introduced and one of them is so called shutter valve. Construction of shutter valve finds its roots in photographic industry where it is commonly used for exposition of photosensitive material for exactly determined period, enforcing very fast operation of mechanism similar like in case of valve train, which makes it suitable to adapt in engine timing system.

1. Introduction

Main function of an internal combustion engine timing system is synchronization of devices responsible for filling of combustion chamber with air-fuel mixture and scavenging it from products of combustion process. Desired phenomena is to charge the cylinder with highest possible amount of air-fuel mixture and get rid of exhaust gases as fast as possible. It appears that commonly used poppet valves reached their limits, that is why new, so called camless systems are becoming increasingly important. Beside more precise timing, such systems enable to switch between work cycles hence improvement in fuel economy. Computer controlled timing also makes possible to get rid of both EGR and throttle valve, what has positive impact on pumping losses [1].

2. Shutter valves

Shutter valve consist of metal casing also named valve body, enclosing two concentric rings among which one is fixed and second one can rotate around center of an aperture. Relative motion of blades is possible thanks to application of iris mechanism.

Depending on construction two main groups of iris valves may be distinguished. First one are valves used as a hydraulic flow regulators in piping industry. Their shape results from necessity of overcoming high pressure differences, be resistant to eventual leaks and cavitation. All of that enforces robust design of components and their precise control, that is why iris valves used in hydraulic systems are characterized by a few relatively thick, non overlapping blades sliding without almost any rotational motion (fig.1b).

Second group of shutter valves are used to control gravity flow discharge of bulk materials. Their design in contrast to heavy duty hydraulic valves, involves multiple overlapping blades having minimal thickness. Greater number of blades results in more circular aperture (fig.1a), which is desirable in case of flow area, however on the other hand, more movable components makes device structure more complex thus hard to manufacture and susceptible to eventual leaks.

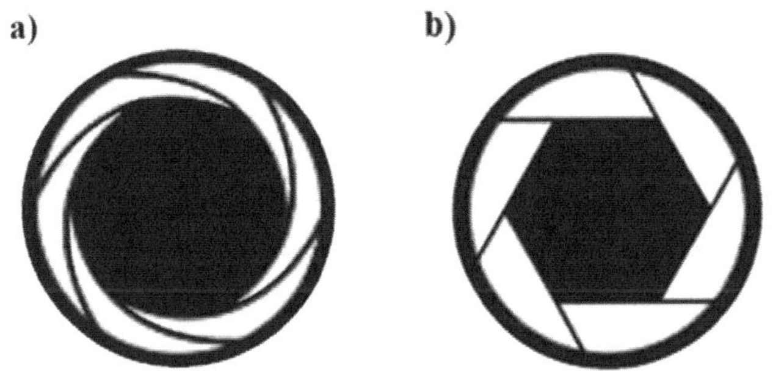

Figure 1. Blades shape impact on aperture [2].

3. Design limitations

One of the most important factors determining valve geometry is the space above combustion chamber. Internal combustion engines produced for a mass scale involves cylindrical shape of piston, which limits the space above it to piston diameter. In order to estimate particular value of space available, a comparison of several engine units were made. Comparison contains data of a spark ignition four cylinder inline engines of displacement class 2000 ccm, which are available in sale on European market.

Table 1. Selected engines data

Manufacturer	Introduction year	Engine code	Displacement (ccm)	Piston diameter (mm)
Toyota[3]	2018	M20A-FKS	1998	80,5
Mazda[4]	2012	PE-VPS	1998	83,5
Volvo[5]	2016	B4204T9	1969	82,0
Mercedes[6]	2017	M260 E20 DE LA	1991	83,0
Volkswagen[7]	2017	EA 888 Gen.3B	1984	82,5

Analysis of data presented in table 1 enabled to assume average cylinder diameter at level of 82mm. Worth mentioning fact for further discussion is that all of considered engine units involves four valves per cylinder layout.

4. Valve geometry determination

First step in valve determination process was valve type selection. Having in mind operational conditions of an engine valve such as its fast operation and space limitation, the overlapping blades layout was chosen.

Thereafter an assumption about equal diameters of both inlet and exhaust valves was made. Initial geometrical analysis indicated that maximal aperture diameter is 41mm in case of two valves per cylinder layout and 34mm for four valves per cylinder. Due to greater total aperture area, four valves per cylinder layout was picked out for further consideration. Subsequently necessity of a sparkplug

IOP Conf. Series: Materials Science and Engineering **614** (2019) 012002 doi:10.1088/1757-899X/614/1/012002

presence was taken into account thus sizing factor sf=0,9 was implemented resulting in aperture diameter decrease up to approximately 31mm.

Next step of analysis involved AutoCAD software in order to estimate blades quantity and size. At this moment it became clear that every blade pivot must be located in the middle of width, at one of blade's ends. In addition rotation angle has to ensure blade to cover the center of valve aperture when closed as it is presented in the figure 2.

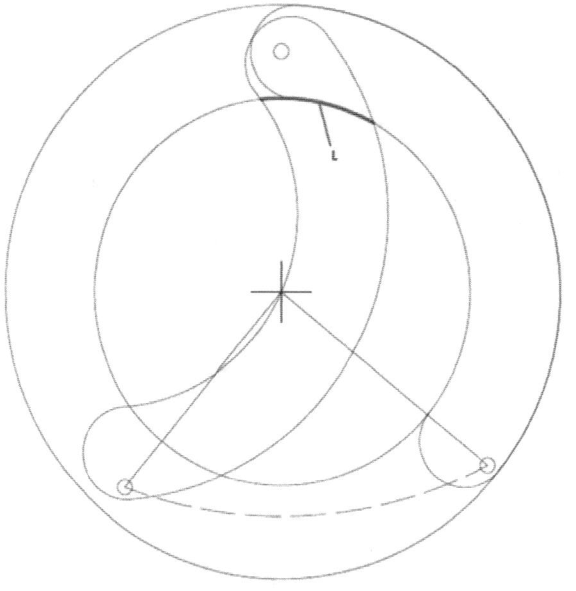

Figure 2. Blade trajectory.

Successive assumption resulted from technological limitations and claimed that number of blades should be as low as possible in order to minimize friction loses [8] and make system relatively simple to manufacture. At this point maximum number of 10 blades was assumed.

The last geometrical requirement applies to outer diameter of shutter valve. Once again limiting factor was diameter of cylinder, which in combination with four valves per cylinder and sizing factor sf=0,9 gave an output of 37mm.

Then a relation between blade width (w) and aperture diameter (d_{in}) was observed. It occurred that with increasing blade (n) width aperture diameter will decrease, however it is not a linear dependence. Other limitations connected with manufacturing enforces blade width being high enough to enable installation of pin at the pivot point. All of those observations resulted in improved design with smaller aperture diameter and lower number of blades, whose width was increased. Results of geometrical analysis are presented in table 2.

Table 2. Geometrical analysis results

d_{in} [mm]	27	25	23
w [mm]	5	6	7
L [mm]	6,23	7,68	9,58
Blade rot. angle [°]	50	48	44
n [-]	14	10	8

Results obtained allowed to make preliminary decision about valve geometry. The most compromise solution between aperture size and number of moving parts seemed to be one with 25mm of internal diameter. In process of continued optimization of a blade shape, their quantity was reduced to 8 pieces. Final layout of valves above combustion chamber is presented in the figure 3.

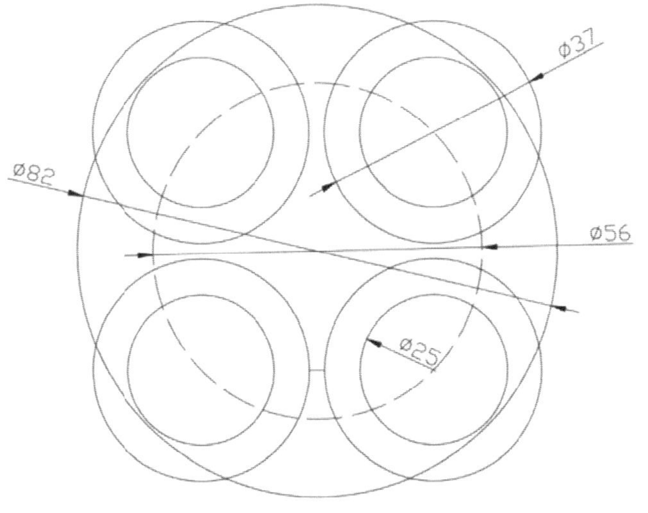

Figure 3. Proposed layout.

5. 3D model

Obtained results enabled to create a 3D model of device with help of Autodesk Inventor software. The model was created taking into account place for sparkplug at the center of combustion chamber and slots for rotating plate arm connecting with an actuator. Due to limited space the decision about manufacturing one casing for two valves was made and blade thickness was set to 0,5 mm.

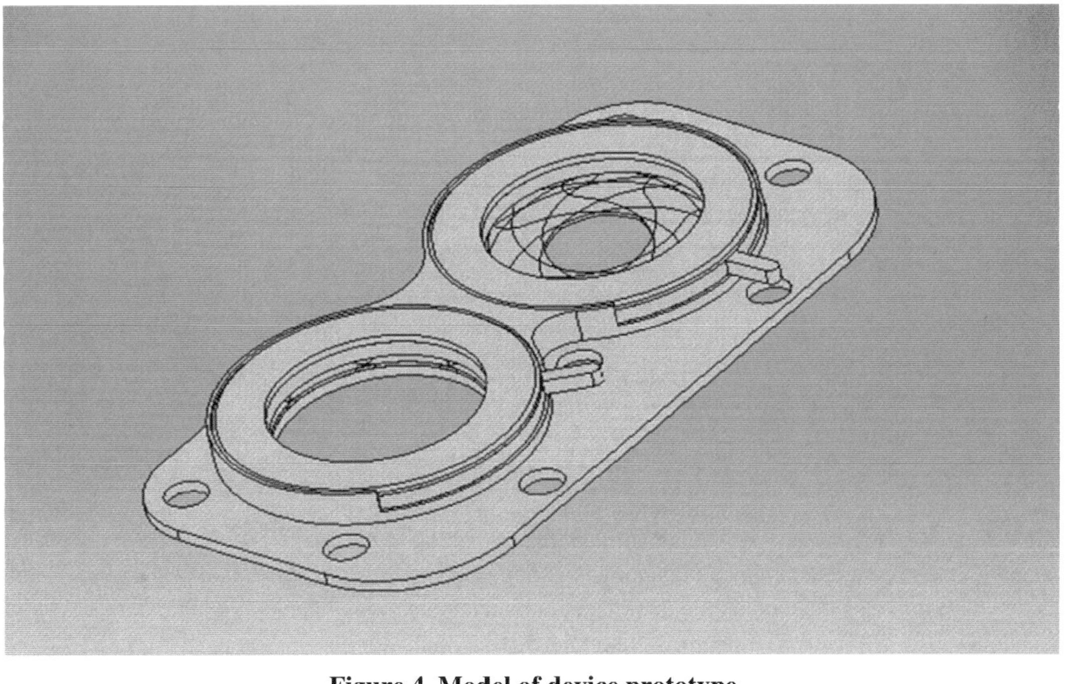

Figure 4. Model of device prototype.

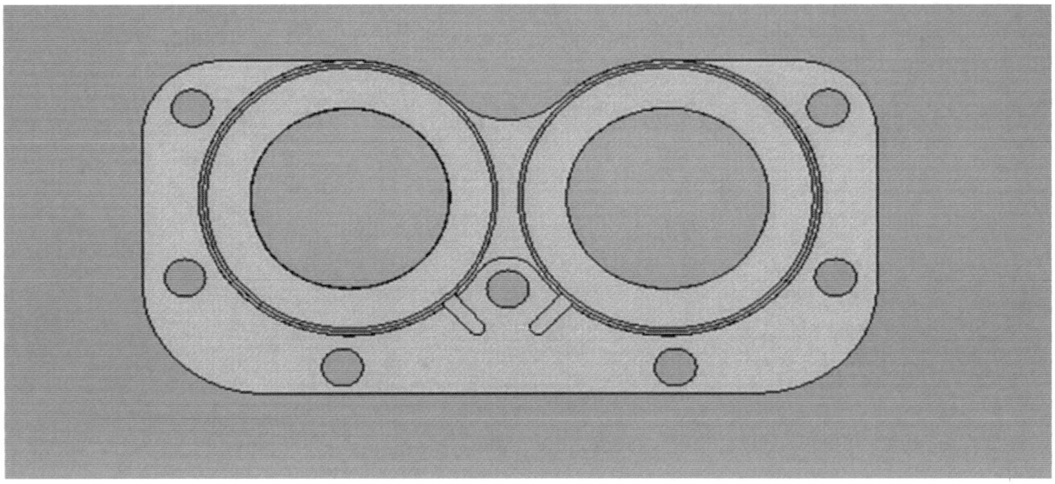

Figure 5. Valves fully opened.

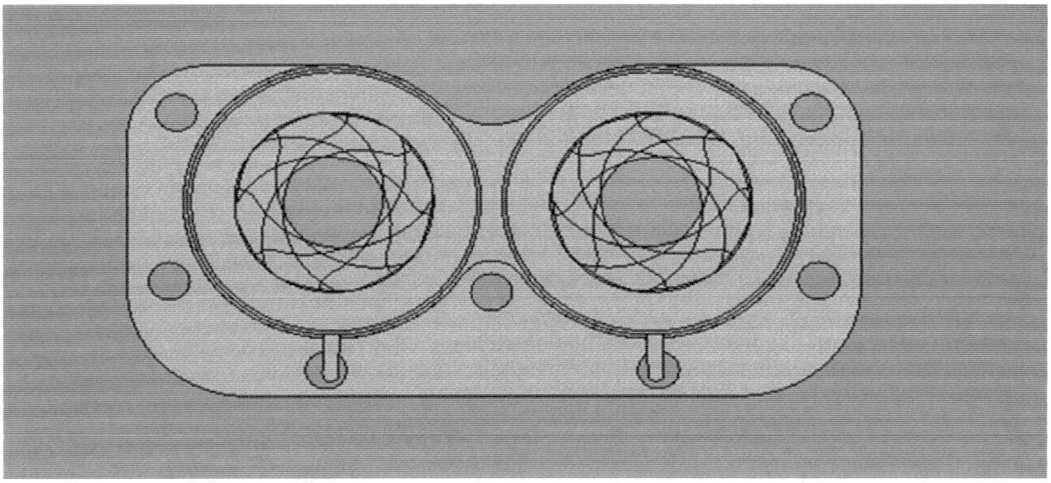

Figure 6. Valves partially closed.

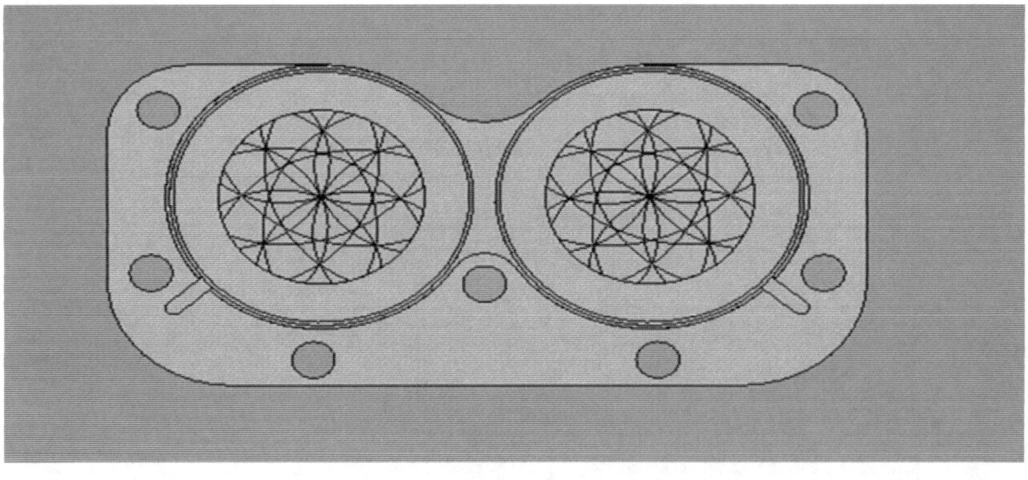

Figure 7. Valves fully closed.

6. Conclusions and discussion

Adaptation of iris mechanism for purposes of timing system seems to generate a lot of benefits. Conventional camshaft located in cylinder head imposes appliance of its propelling system, which generates losses connected with friction and inertia. Decrease of inertia forces should allow to employ electromagnetic actuators in order to control valve timing similar like in case of Free Valve system [9].

Shutter valve timing system reveals many theoretical advantages over conventional or even other camless systems due to lack of conventional poppet valve. Such state of matter makes whole system more compact in construction and what is most important enable to extend internal combustion engine range of operation. Thanks to implementation of shutter valves it should be possible to adjust air-fuel ratio, temperature of combustion, residual exhaust gases contribution or even change of engine work cycle. Biggest problem at this moment seems to be selection of materials able to resist rapid pressure fluctuation, complex thermal loads and friction [10].

Following publication should be treated as a first step in revolutionary solution implementation. Further work should be focused on manufacturing of prototype and development of control system. Later on improvement in valve geometry should be done taking into account cooling channels and mass production technologies.

References

[1] Matla J. 2019 *Design of Timing System Using Shutter Valves in an Internal Combustion Engine* (Wrocław: Wrocław University of Science and Technology - Diploma Thesis).
[2] https://www.canstockphoto.com/camera-shutter-icons-set-black-simple-36316166.html, accessed on 25.06.2019.
[3] https://toyota-club.net/files/faq/18-03-20_faq_df_r4_eng.htm, accessed on 26.06.2019.
[4] https://en.wikipedia.org/wiki/SkyActiv#SkyActiv-G, accessed on 26.06.2019.
[5] http://australiancar.reviews/Volvo_VEP4_VEA_Engine.php, accessed on 26.06.2019.
[6] https://en.wikipedia.org/wiki/Mercedes-Benz_M260/M264_engine, accessed on 26.06.2019.
[7] https://www.greencarcongress.com/2017/05/20170523-tiguan.html, accessed on 26.06.2019.
[8] Siczek K.J. 2016 *Tribological processes in the valve train systems with lightweight valves* Elsevier.
[9] Gundersen Ø. 2009 *Free Valve Technology* (Stockholm: KTH Industral Engineering and Management – Diploma thesis).
[10] Sroka Z. J. 2012 *Some aspects of thermal load and operating indexes after downsizing for internal combustion engine* Journal of Thermal Analysis and Calorimetry, Volume 110, Issue 1, p. 51–58.

Method for determining the parameters of injection of Common Rail injectors

Sergey G Belchev

Department of Automotive Engineering and Technologies, Faculty of Mechanical Engineering and Technologies, Technical University - Varna, 1 Studentska Str., city of Varna, Bulgaria

sergtu@abv.bg

Abstract. The flow rate through the holes of the nozzle at constant pressure and steady flow is an important feature of electronically controlled injectors. The flow rates at different pressure of injection determine the main characteristic of Common Rail injectors. In scientific researches it is necessary to determine this flow rate experimentally. This requires specially equipped experimental installations and expensive measuring devices. The article presents a method for determining the steady flow rate through the holes of the nozzles using a standard test bench for Common Rail injectors. The method is based on the assumption that, all else being equal, non-steady phenomena at the beginning and end of the injection and displacement of injection to the control pulse are identical at different injection timing. Under this conditions we can easily determine the steady flow rate through the holes of the nozzle.

1. Introduction

In modern diesel engines with Common Rail fuel system the injected fuel quantity is controlled by injectors. Injected fuel quantity is implemented, setting the injection timing at a given injection pressure. The injectors are electronically controlled and the time during which they are open and inject fuel is determined by the duration of control electronic pulse.

The magnitude of fuel flow rate through the holes of the nozzles at different injection pressures is an important feature of these injectors. There are three different stages of injection with longer duration. The first opening stage of injector is characterised by non-steady flow with increasing flow rate. The third closing stage of injector is characterised by non-steady flow with decreasing flow rate. During the second stage, when the nozzle needle is fixed and fully open, flow of leakage with constant flow rate is established. Precisely this constant flow rate determines the injector feature and is needed to design and test Common Rail injectors. It is determined theoretically in design calculations and experimentally in the presence of a real injector. Experimental calculation, however, faces many difficulties. Expensive devices and complex experimental installations are needed to determine the steady flow rate through the holes of the nozzles.

The objective of this work is to determine this important feature of Common Rail injectors, using standard test benches for such injectors. In addition, the test results need to be processed, using simple mathematical device in order to avoid any additional inaccuracies in the obtained flow rates through the injectors.

2. Prerequisites and Basic Concepts

Injected fuel quantity of Common Rail electromagnetic injectors is controlled with the help of electronic pulse of a certain nature and duration. The electronic pulse nature is known [1]. It is determined by the intensity of the current passing through the solenoid valve and consists of several phases (a, b, c, d, e), figure 1. The pulse nature is determined by the need for opening the valve quickly and for closing it immediately after interruption of the pulse. The letter "F" on the figure shows the injection end, the letter "t" shows the control pulse duration and a "t_{inj}" shows the injection timing.

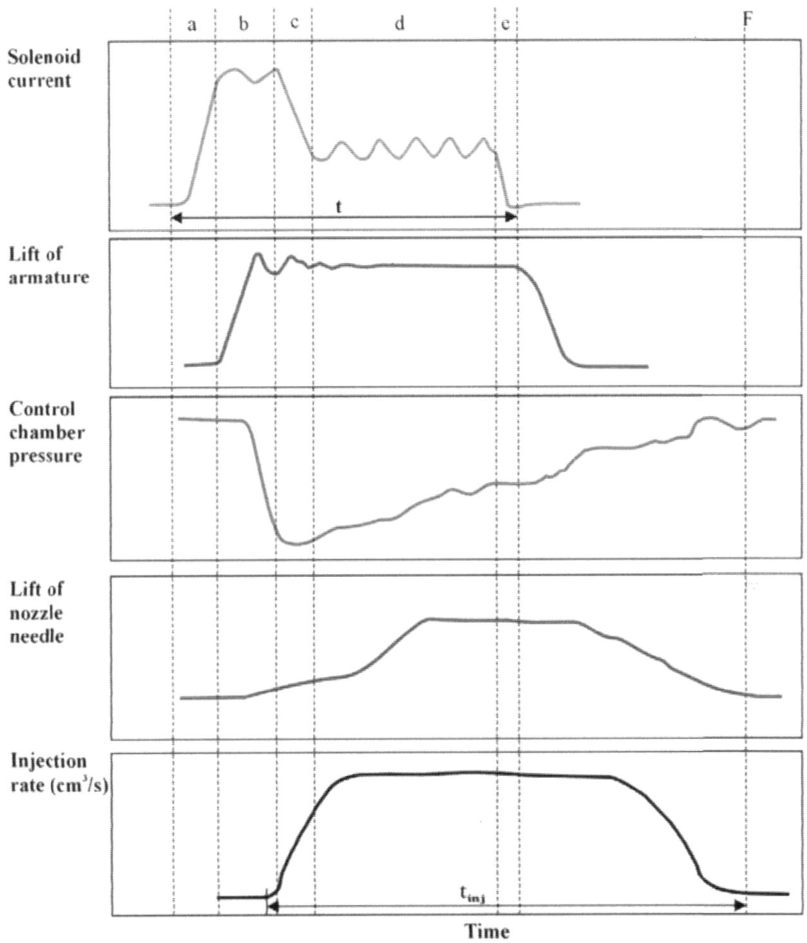

Figure 1. Stages of a traditional injection process.

In fact, the movement of solenoid valve armature and related ball is delayed compared to that of the electronic pulse [1]. The valve opens later and closes later. The pressure cycle in valve control chamber also lags behind the valve movement, the nature of this cycle not copying the valve movement but determined by it [2]. The pressure cycle in valve control chamber is a direct reason to drive the valve control plunger connected with the nozzle needle. In general, the needle movement due to the changing pressure in valve control chamber lags behind that of the control pulse, figure 1. When moving, the nozzle needle opens the nozzle openings and determines the beginning, duration, and end of the fuel injection process [2], [3].

At the beginning of needle lifting, while the clear opening in the nozzle needle seat taper is less than its openings, the fuel flow rate increases sharply and reaches certain constant value at a given needle position. The next needle movement up to its final position does not change the magnitude of fuel flow rate. A period of detected leakage at constant flow rate follows. During the needle closing, fuel flow rate remains constant for some time, and then it sharply reduces until the final stop of the needle movement, figure 1. Except for the start and end phases with variable flow rate, the average constant value of the fuel flow rate during the injection should be determined.

In case of longer injection and control pulse duration, and nature of change in fuel flow rate it may be concluded that the nature of change in fuel flow rate does not depend on the pulse duration. The nature of fuel flow rate initial increase and late injection do not depend on the control pulse duration. There are many factors that influence the nature of initial injection, but if they are the same for the same injector and will therefore not change the beginning of injection. The same applies to the end of injection. In case of sufficient injection timing, the phenomena accompanying needle opening subside and a steady flow with constant flow rate is established. It is therefore assumed that the phenomena accompanying the nozzle needle closing and nozzle late closure do not depend on the length of the previous period of detected leakage as long as this duration is sufficiently large.

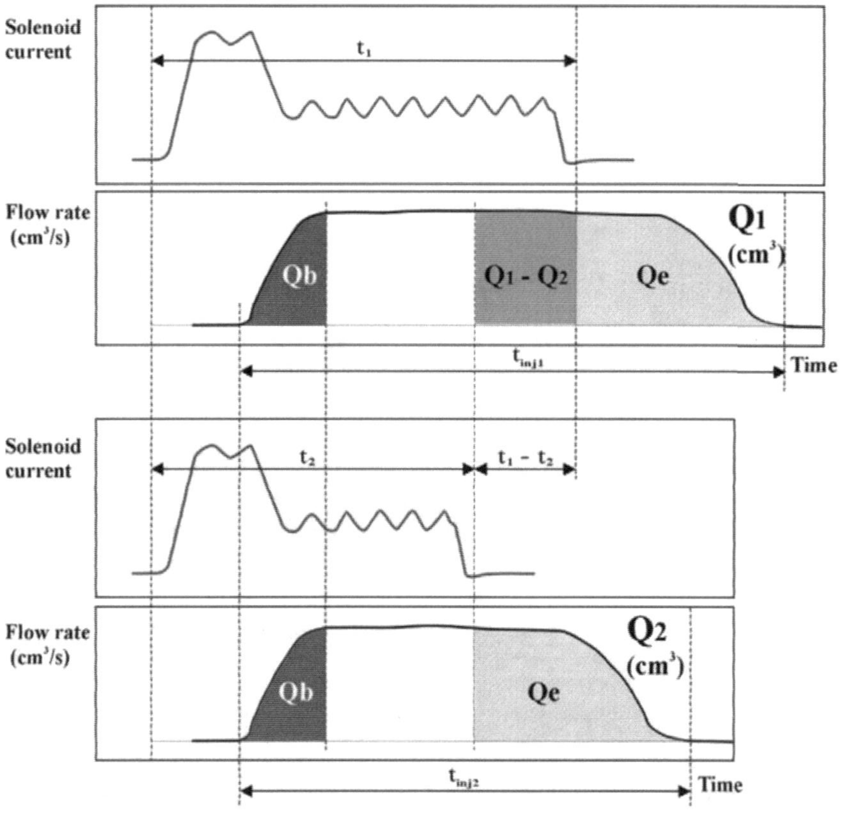

Figure 2. Method for determining the steady flow rate G_{Vst}.

The considerations outlined above lead to the conclusion that the increased sufficiently long duration of control pulse results in increase in the injection timing with the same value, which the control pulse is increased with The increased injected fuel is at the expense of the increased time of detected leakage at constant flow rate. This is illustrated in figure 2. On the figure, *Qb* indicates the quantity of fuel injected during the initial lifting of nozzle needle. *Qe* is the quantity of fuel injected after the control pulse is interrupted, which includes non-steady mode in the needle closing. Q_1 and Q_2

indicate the total quantity of fuel injected at control pulse duration of t_1 and t_2, respectively. These fuel amounts are proportional to the area under the fuel flow change curves. The injection timing at a different control pulse is indicated by t_{inj1} and t_{inj2}, respectively.

Therefore, it could be argued that:

$$t_{inj1} - t_{inj2} = t_1 - t_2 \tag{1}$$

As outlined in the considerations above and as shown in figure 2, the simple dependence for calculating G_{Vst} - fuel flow rate value - may be determined at a steady flow of leakage:

$$G_{Vst} = \frac{Q1-Q2}{t_1-t_2},\ cm^3/s \tag{2}$$

With the help of formula (2), it is possible to calculate the value of the constant flow rate through the holes of the nozzles for a real injector. It is only needed to experimentally determine the injected fuel for this injector at two different durations of the control pulse, all else being equal. Injectors can be tested on a standard test bench for Common Rail injectors which provides sufficient accuracy of the measured fuel quantities. The expansion of experiment will allow calculation of the flow rate at different pressures and determination of the real injector feature.

3. Experimental Studies

The experiments demonstrating the above statements are made on a standard test bench for Common Rail injectors CRI-NTR816E. The elements of the system for pressure generation, measurement and control are produces by the company BOSCH. The measurement is made by Cobolt system, with accuracy of the measurement of 0.1 mm^3. The studies were conducted with a BOSCH working injector having number 0445110044.

The injected fuel quantities are measured at three different pressure levels in the fuel rail - 135MPa, 80MPa and 25MPa. Four different control pulse durations - 1800μs, 1500μs, 1200μs and 900μs - are obtained for each pressure level. Three measurements are made for each experiment points and the results are averaged.

$$Q_m = \frac{Q_1+Q_2+Q_3}{3},\ mm^3 \tag{3}$$

$$G_{V'} = \frac{1000Q_m}{t},\ cm^3/s \tag{4}$$

Table 1 shows the experiment results. The last column shows the flow rate value calculated with the help of formula (4). It is not real, since the injection time is not equal to the control pulse duration but can be used for comparison.

Formula (2) calculates G_{Vst} - flow rate values - at detected leakage using various pulse duration combinations. The results obtained for the three pressure levels in the fuel rail are presented in table 2. The first row of the table shows the time values in (μs) which are used to make the calculations.

4. Analysis of the Experimental Results

As apparent from table 2, the various combinations of pulse durations lead to different G_{Vst} flow rate values. The average values are closest to the results obtained at selected values t_1=1800μs and t_2=900μs, (1800-900). This is the logical consequence of the fact that the difference between the two times is the highest which minimises the error of temporary disturbances during injection at a steady flow. If the difference in the duration of pulses used is smaller, small deviation from the steady flow permanent nature resulting from wave phenomena may be expected.

IOP Conf. Series: Materials Science and Engineering **614** (2019) 012003 doi:10.1088/1757-899X/614/1/012003

Table 1. Experiment results

	t μs	*Q1* mm^3	*Q2* mm^3	*Q3* mm^3	*Qm* mm^3	*Gv'* cm^3/s
135MPa	1800	86.5	86.9	86.6	86.67	48.15
	1500	75.9	76.1	76.4	76.13	50.75
	1200	60.6	61.2	62.7	61.50	51.25
	900	48,0	48.3	48.4	48.23	53.59
80MPa	1800	65.4	65	65	65.13	36.18
	1500	55.1	55.8	55.8	55.57	37.05
	1200	44.9	45.3	45.4	45.20	37.67
	900	34.7	35.2	35.3	35.07	38.97
25MPa	1800	32.4	32.7	32.8	32.63	18.13
	1500	26.9	26.9	27.8	27.20	18.13
	1200	18.3	18.5	19.4	18.73	15.61
	900	11.6	11.8	11.8	11.73	13.03

Table 2. Calculated G_{VSt} flow rate values at steady leakage flow

P_{inj}	1800-1500	1800-1200	1800-900	1500-1200	1500-900	1200-900
135MPa	35.11	41.94	42.70	48.78	46.50	44.22
80MPa	31.89	33.22	33.41	34.56	34.17	33.78
25MPa	18.11	23.17	23.22	28.22	25.78	23.33

To obtain adequate results with a small error, the duration t_I should not be less than 900μs. When testing a specific injector, an experiment can be made by increasing the injection times and the interval between them to detect the smallest values after which the increase does not result in a significant change in the calculated values of the established G_{VSt} flow rate.

When comparing the relative indicator *Gv'* with the calculated G_{VSt} values, it can be concluded that at higher injection pressures (135MPa, 80MPa) *Gv'* is higher than G_{VSt}. At a low pressure of 25MPa, *Gv '* is lower than G_{VSt}. The likely explanation of this fact is:

- At high injection pressure, the needle response time during its opening is less, due to the higher absolute pressure drop in the valve control chamber and vice versa - the needle closing takes more time due to the slower pressure equalization in the valve control chamber. This increases the injection time and it may exceed the control pulse time;
- At low injection pressure, the needle response time is increased and the delay of initial injection is also increased compared to that of the initial pulse. On the other hand, the needle closes faster due to the faster pressure equalization in the valve control chamber. This reduces the injection time to the control pulse duration.

Figure 3 illustrates the change in the nature of injection at equal times of the control pulse and different injection pressures.

Figure 3 shows the injection parameters at low pressure (lp) and at high pressure (hp). It also shows approximately the average injection flow rate calculated by formula (5), where the fuel injected is relative to the injection time.

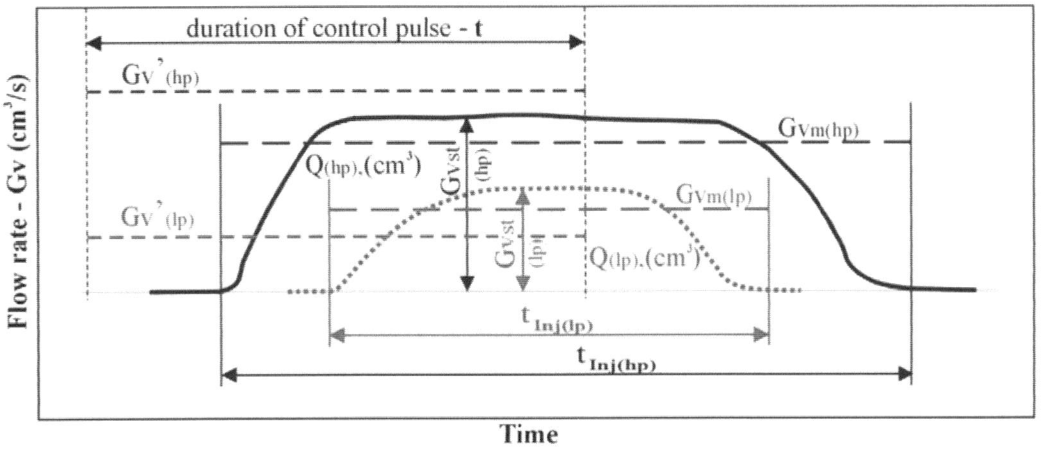

Figure 3. Injection parameters at different pressures.

$$Q_{Vm} = \frac{Q}{t_{inj}}, \ cm^3/s \tag{5}$$

This average flow rate value cannot be calculated using the presented methodology, since the injection time is unknown. However, based on the methodology, information on the relative injection time compared to the pulse duration can be obtained:

- if Gv' is higher than G_{Vst}, then t_{inj} is higher than t - at a relatively high injection pressure;
- if Gv' is lower than G_{Vst}, then t_{inj} is lower than t - at a relatively lo injection pressure.

There is obviously a pressure where Gv' is equal to G_{Vst} and therefore t_{inj} is equal to t. It is possible to find out this pressure at which the above values are equal with a series of experiments.

5. Conclusion

In the presence of a test bench for Common Rail injectors, and with the help of the presented methodology, it is possible to easily and accurately determine the flow rate through the holes of the nozzle at detected leakage flow.

The pulse duration should be relatively large, more than 900μs, to ensure a period of detected leakage.

In case of greater difference in the duration of pulses used in the calculation, more precise results are obtained.

The deviations of the obtained debit rates are greater at very high pressure levels. In such cases pulses with greater difference between them are needed.

Based on the methodology, additional conclusions on the nature and duration of fuel injection can be made.

6. References

[1] Robert Bosch GmbH 2004 *Diesel-Engine Management, 4th Edition* (Stuttgart) p 504.
[2] Xiong J and Gu H 2017 An Intelligent Dual-Voltage Driving Method and Circuit For a Common Rail Injector for Heavy-Duty Diesel Engines *J. IEEE Access* **5** 27681-27689.
[3] Dongiovanni C and Coppo M 2012 Accurate Modelling of an Injector for Common Rail Systems *Fuel Injection* Siano D (Rijeka: InTech) p 262.

Acknowledgments
This work was completed thanks to the generous cooperation of Diesel Service Varna, which provided its equipment, benches and injectors for research.

IOP Conf. Series: Materials Science and Engineering **614** (2019) 012004 doi:10.1088/1757-899X/614/1/012004

Method for assessing the adequacy of a crankshaft strain state model

S Belchev[1,2], K Todorov[1], G Chekelov[1] and D Petkov[1]

[1] Department of Automotive Engineering and Technologies, Faculty of Mechanical Engineering and Technologies, Technical University - Varna, 1 Studentska Str., city of Varna, Bulgaria

[2] E-mail: sergtu@abv.bg

Abstract. Using the Finite Element Method (FEM) in the SolidWorks software environment, the strain crankshaft state is modelled. The assessment of the adequacy of the model by experimental crankshaft in a running engine is complex and is possible for a small number of control points. The method shown proposes to replace the variable crankshaft strains in motion by examining a static strain state of the crankshaft. The results obtained show a high degree of adequacy of the model. The differences can be minimised by specifying the physical properties of the material used in the simulation, taking into account the elasticity of the supports.

1. Introduction

In the design phase of internal combustion engines, models are always used to determine various phenomena and processes which occur in their operation. The study of stress and strain state of the engine's main elements is inconceivable without the application of Finite Element Method (FEM). It is applied in various software environments, one of which is SolidWorks [1]. In this software environment, a model of diesel engine crankshaft has been created [2]. The results obtained by the model give an opportunity to assess the impact of small changes in the crankshaft shape on its strength and its stress and strain state.

The created model adequacy is assessed experimentally. This assessment in needed to prove the relevance of the results obtained through the model. The adequacy is usually proven by measuring the stress of prepared crankshaft placed under loading conditions close to the real ones. The complexity of this task results in compromises and large inaccuracies of the experiment and therefore the safety of the results obtained in the measurement is small. On the other hand, the accuracy of mechanical characteristics laid down in the model and boundary loading and attachment conditions are crucial to the accuracy of the results obtained from the model studies. In fact, it is more important to set the model and its parameters so that it corresponds to reality. It is not necessarily to do this by testing a crankshaft under real operating conditions.

The aim of the presented work is to test a method for assessing and specifying the crankshaft model by static shaft loading and measuring deformations in a sufficient number of points. The degree of coincidence of the measured deformations with those obtained in the modelling is sufficient criterion for adequacy of the developed crankshaft model.

2. Experimental test and model

Crankshaft mounted on a lathe is used in the presented method. The shaft front end is fixed to the lathe chuck and the rear end is fixed to the lathe with the help of alignment device. Thus, the shaft can rotate but can not change its axis of rotation, figure 1. A lever is attached to the crankshaft rear end by means of transition flange.

Figure 1. Fastening of the crankshaft to a lathe.

At the end of the web, different weights can be attached. Sufficiently rigid support is placed under crank pin 1 (which is the closest to shaft front end) so that the shaft can not be turned over, figure 2. The load applied to the lever, combined with the support under crank pin, creates torsion of the crank pin crankshaft. The shaft deformations and corresponding displacements of certain points on its surface can be measured with the help of dial gauge.

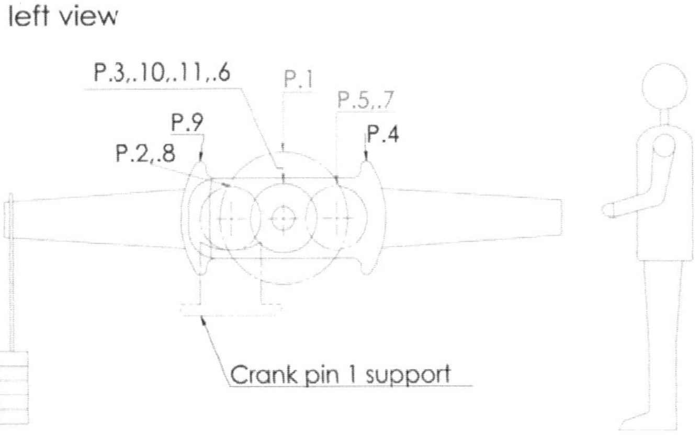

Figure 2. Scheme of fastening and load

Figure 3 shows the points of measurement. These are the central areas on the upper surface of all crank pins (P.2, P.5, P.7, P.8), two of the counterweights (P.4, P.9) and the main crank pins (P.3 , P.6, P10, P.11). Measurement is also performed on the flange at the crankshaft rear end on which the web is attached (P.1). The accuracy of displacement measurement is 0.001mm.

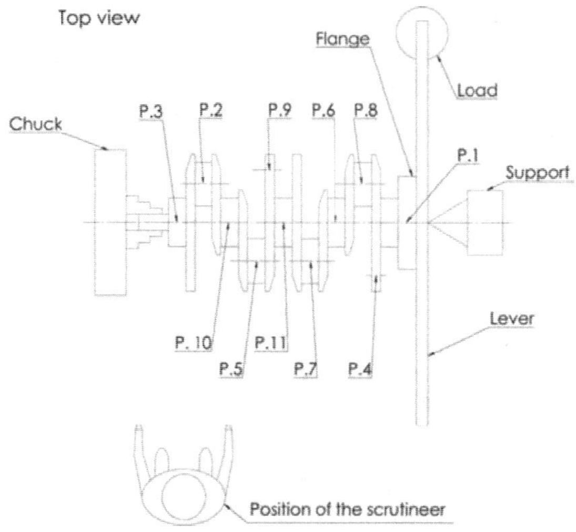

Figure 3. Top view - points of measurement.

Three-dimensional model is created for the actual crankshaft and additional elements in the SolidWorks environment. All details of the shaft shape, flange, web and support are reflected. Figure 4 shows the connection between the crankshaft rear end and web through the transition flange. The three-dimensional image of the model copies precisely the real elements. Figure 5 shows the crankshaft and type of support under the first crank pin.

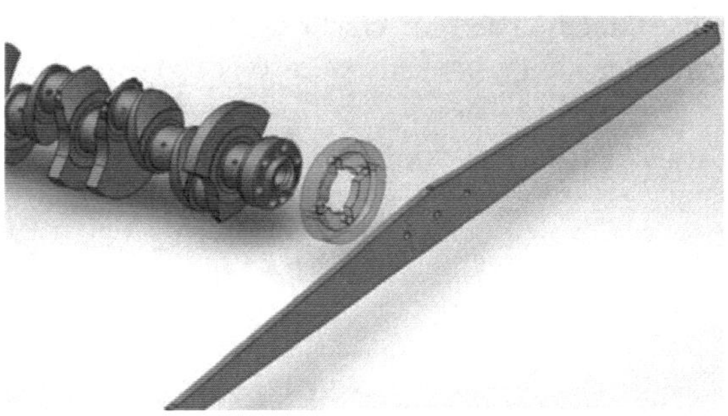

Figure 4. Rear end crankshaft, flange and lever

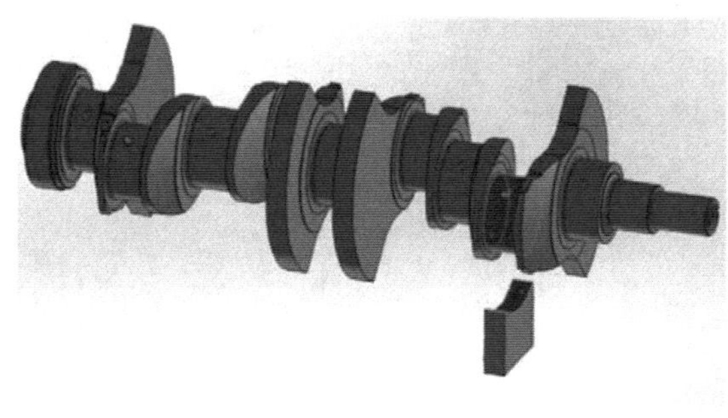

Figure 5. Support under the first crank pin

The boundary fastening conditions are as follows:

- at the crankshaft front end bearing support and axis fixing are implemented, figure 6;
- the support under the first crank pin is permanently fixed in all directions; it in turn restricts the shaft rotation, figure 6;
- at the crankshaft rear end bearing support is provided in the web opening - contact with the lathe alignment device is made there, figure 7;
- loading is set on a small platform at the web end, figure 7.

The initial version of the model with absolutely rigid supports is not very reliable. Therefore, after measuring the deformations in the supports and calculating the supporting reactions, the support stiffness was determined and set in the model, table 1. Under these fastening conditions, the tests in SolidWorks environment have been performed, using elastic supports.

IOP Conf. Series: Materials Science and Engineering **614** (2019) 012004 doi:10.1088/1757-899X/614/1/012004

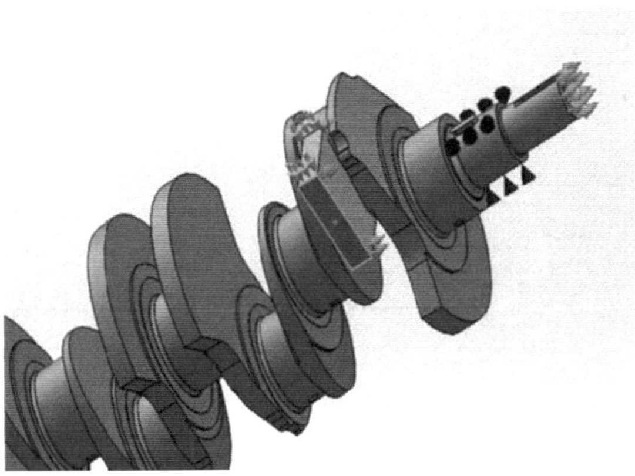

Figure 6. Fastening the front end of the crank-shaft and support under the first crank pin

Figure 7. Fastening the rear end of the crank-shaft and loading force

Table 1. Reactions and stiffness of the supports at maximum load 824N.

Support - point №	Point 1	Point 2	Point 3
Reaction of supports (N)	-445.77	+8392.07	-7122.3
Stiffness of supports (MN/m)	0,9103	275,252	36,928

3. Experimental results

Deformations in the eleven points shown are measured consequently three times and the average values of the three measurements are accepted. The system was initially loaded with approximately 40N to remove the clearance. The values at this loading are accepted as initial ones and the instrumentation is reset. The load is then increased to approximately 200N to reach the maximum load of 824N and is reduced by the same step to the initial level.

Table 2. Average values of measured elastic strains.

Load	Point 1	Point 2	Point 3	Point 4	Point 5	Point 6	Point 7				Point 8	Point 9	Point10	Point11
(N)	average	average	average	average	average	average	meas.1	meas.2	meas.3	average	average	average	average	average
39,2	0	0	0	0	0	0	0	0	0	0	0	0	0	0
235,4	0,0033	-0,0257	0,0800	0,0557	0,4180	0,1200	0,401	0,382	0,385	0,3893	-0,1493	-0,3167	0,1800	0,1617
431,6	0,0207	-0,0330	0,1783	0,1150	0,7750	0,2333	0,745	0,724	0,729	0,7327	-0,2680	-0,5400	0,3550	0,3217
627,8	0,0433	-0,0377	0,2210	0,1683	0,9807	0,3067	0,933	0,922	0,927	0,9273	-0,3247	-0,7300	0,4600	0,4153
824	0,0567	-0,0427	0,2727	0,2050	1,1767	0,3667	1,115	1,093	1,100	1,1027	-0,3850	-0,8883	0,5467	0,4977
627,8	0,0667	-0,0433	0,2817	0,2133	1,0357	0,3417	0,972	0,951	0,957	0,9600	-0,3253	-0,8417	0,4883	0,4363
431,6	0,0623	-0,0433	0,2467	0,2067	0,8497	0,2867	0,794	0,781	0,785	0,7867	-0,2637	-0,6800	0,4083	0,3570
235,4	0,0483	-0,0400	0,1377	0,1167	0,4777	0,1617	0,448	0,431	0,438	0,4390	-0,0520	-0,4117	0,2200	0,2047
39,2	0,0117	-0,0013	0,0000	-0,0067	0,0047	0,0133	0,016	-0,007	0,000	0,0030	0,0017	-0,0383	0,0200	0,0127

Table 2 shows the average values of the measured deformations at all points. For example, the values of the three measurements are shown only in point 7. For illustration, figure 8 shows the changes in the deformations of the main journals at the change of loading.

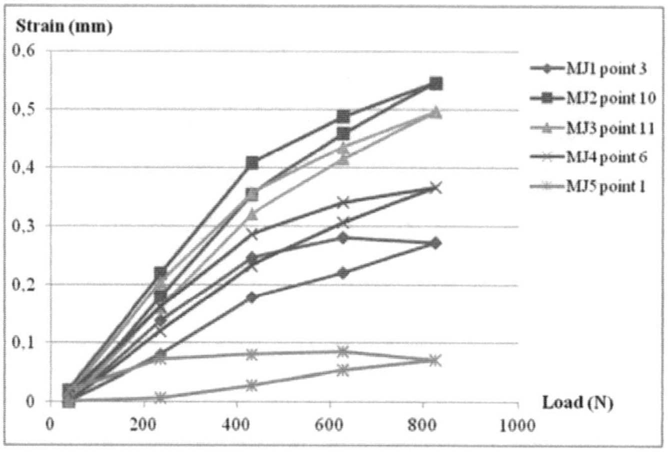

Figure 8. Measured strains of the main journals(MJ) of the crankshaft.

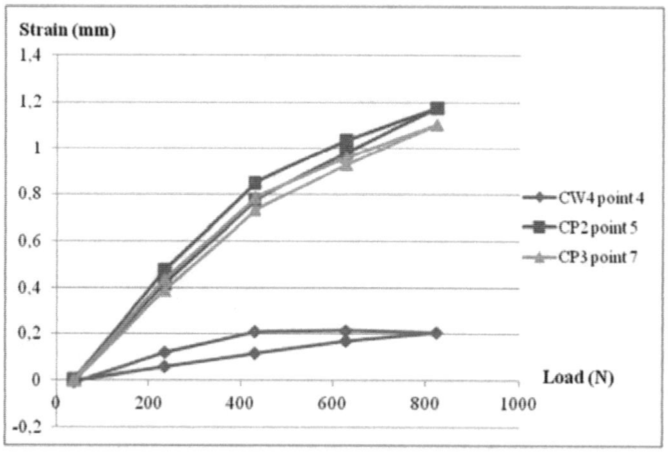

Figure 9. Measured strains of the crank pins 2 and 3(CP) and counterweight 4(CW4) of the crankshaft.

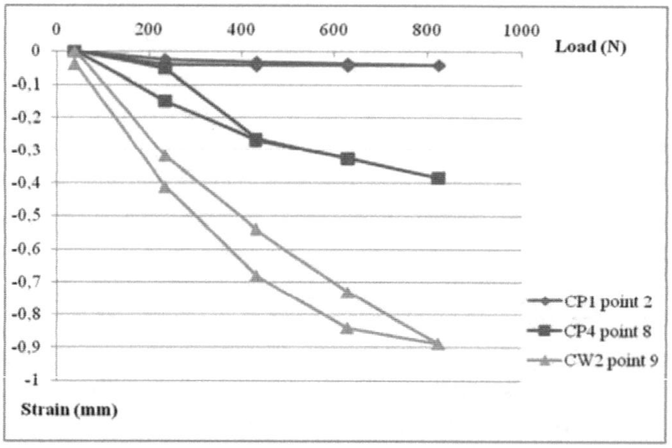

Figure 10. Measured strains of the crank pins 1 and 4(CP) and counterweight 2(CW2) of the crankshaft.

Figure 9 shows the deformations of points from counterweight 4 and crank pins 2 and 3 that move in the direction chosen as positive. Figure 10 shows the negative displacements of the points from counterweight 2 and crank pins 1 and 4.

4. Adequacy of the model

The same loads as those used in the real experiment are applied to the crankshaft model described above. Two tests were made - test with absolutely rigid supports at points 1, 2 and 3 and test with the same supports which elasticity is defined in table 1. Table 3 shows the results of the numerical studies of the displacements of all points with rigid and elastic supports and comparison with the experimental results. The values shown are in (mm).

Table 3. Comparison of results from numerical studies and experiment.

Type of study	Point 1	Point 2	Point 3	Point 4	Point 5	Point 6	Point 7	Point 8	Point 9	Point 10	Point 11
rigid supports - *rs*	0,0271	-0,0141	0,1063	0,3479	0,5684	0,2554	0,5644	-0,1585	-0,0513	0,2707	0,2839
elastic supports - *es*	0,0580	-0,0425	0,2819	0,3697	0,7923	0,3169	0,7619	-0,2649	-0,1870	0,3556	0,3315
experimental - *e*	0,0567	-0,0427	0,2727	0,2050	1,1767	0,3667	1,1027	-0,3850	-0,8883	0,5467	0,4977
$\Delta = e - es$ (mm)	-0,0013	-0,0002	-0,0092	-0,1647	0,3844	0,0498	0,3408	-0,1201	-0,7013	0,1911	0,1662
$\Delta\% = 100\Delta/e$ (%)	-2,3	0,4	-3,4	-80,3	32,7	13,6	30,9	31,2	78,9	35,0	33,4

The first row in table 3 shows the results of the deformations obtained through the model in SolidWorks environment with absolutely rigid supports in points 1, 2 and 3. The second row shows the results in the use of elastic supports with a stiffness set out in table 1. The next row shows the experimental results. Below are the differences between the experimental results and elastic support study, with the absolute values of the differences and their relative value in percentage being shown.

The deformation direction of the model and that of the real crankshaft coincide in all control points. Larger deviations at points 4 and 9 are noticeable. These are counterweight points. They are located at the greatest distance from the shaft axis and the probability of a measurement error is greatest. In addition, they fall outside of the field lines through the crankshaft under this loading scheme and are not important to the overall strain state of the shaft. The deviations of the support points are negligible. The deviations of the remaining points are within the limits and can be reduced in case of any change in the elastic characteristics of the material set in the crankshaft model.

One of the objectives of the proposed method is to adjust the model parameters so as to achieve maximum identity with the real crankshaft. Then crankshaft studies can be made with the specified parameters of the model in the set operating and loading conditions.

5. Conclusion

The proposed method of specifying the crankshaft model and assessing its adequacy is applicable to crankshafts of automobile engines and to modelling in SolidWorks environment.

The approach in the proposed method is universal and it can be applied to models created with other software that uses the Finite Element Method.

Setting the numerical model and specifying the material characteristics set therein is perfectly possible with high precision through the proposed method.

6. References

[1] Karavasilev O, Kozhuharov M, Dimitrov N, Grigorov B and Mitrev R 2012 *SolidWorks-Basic modeling and drawings* (Sofia: TechnoLogica) p 600.

[2] Belchev S, Sarov O and Hristov R 2015 Study of deformation and strain state of the crankshaft *J. EcoVarna* **22** 461.

Acknowledgments
This work was completed thanks to the funding and equipment provided by Technical University - city of Varna.

IOP Conf. Series: Materials Science and Engineering **614** (2019) 012005 doi:10.1088/1757-899X/614/1/012005

Study on the classification of actions aimed to reduce the fuel consumption of road vehicles in service

L Luca[1*] , M M Pasare[2], R Dimitrov[3], T O Luca[4]

[1]University Constantin Brancusi of Targu-Jiu, Faculty of Engineering, Str. Eroilor, No.30, Targu-Jiu, Romania
[2]University Constantin Brancusi of Targu-Jiu, Faculty of Engineering, Str. Eroilor, No.30, Targu-Jiu, Romania
[3] Technical University of Varna, Department of Automotive Engineering and Technologies , Faculty of Mechanical Engineering and Technologies, Studentska str. №1Varna, Bulgaria
[4]NURVIL SRL Targu-Jiu, Str. Tismana, no.100, Targu-Jiu, Romania

[*]Corresponding author: lylyanaluca@yahoo.com

Abstract: This paper addresses the major global problem of reducing fuel consumption in the transport sector. The authors present the results of a study on the use of a quality management tool - tree diagram. The actions aimed at reducing the fuel consumption of the road vehicles in operation have been identified and classified. A three-level tree diagram was developed to answer the question: How can fuel consumption be reduced? The chart provides a complete and suggestive picture of the actions to be taken to solve the proposed problem. The actions have been grouped into three corresponding categories for three specific objectives: the acquisition of a vehicle suitable for the needs; proper use of the vehicle (eco-driving); proper maintenance. The authors of the paper believe that the tree diagram obtained is a good tool for informing and raising the driver's awareness of his role in reducing fuel consumption.

1. Introduction

Nowadays, automobiles are a very useful means of transport for comfortable transport and offer freedom of movement over short or long distances to solve work problems or for leisure travelers. The road transport sector has an upward trend influenced by human society which is in continuous development. This has led to an increase in the total number of means of transport and the increase in the total distance covered by motor vehicles. The final effect is also a negative one due to the excessive increase in fuel consumption. Big fuel consumption is a topical issue, as oil reserves are exhaustible within a relatively short period of time. Also, the automobiles by polluant emissions have negative effects on the environment and human health. Carbon dioxide, emitted by cars in a quantity proportional to fuel consumption, is responsible for the global warming of the planet and negatively affects sustainable development. Another issue for transport companies is related to the cost of purchasing fuel that has increased much in recent years. They are directly interested in reducing fuel consumption.

Reduce fuel consumption is a research theme at world level, which concerns in particular the engineers in the field of construction of cars. Their attention was heading toward more and more constructive variants with low power consumption of fuel.

In the first part of the 21st century, there was an increased concern for the construction of new hybrid and electric road vehicles. The number of new hybrid cars and electric vehicles is growing, but the infrastructure for electric vehicles is still underdeveloped. Therefore, reducing the fuel consumption of road vehicles in service remains a major concern.

Specialty literature offers a lot of papers dealing with fuel consumption. In the following, we present for examples, only a few of the papers dealing with this issue. The paper [1] presents a study on the use of a new device attached to the engine to reduce fuel consumption. In the papers [2, 3] are proposed models based on the Genetic Algorithm (GA), to reduce and estimate the fuel consumption based on the optimization of transport parameters. Also based on the Genetic Algorithm, a management scheme for efficient fuel consumption was established [4]. The paper [5] presents an analysis of the influence of driver training programs on reducing fuel consumption. The paper [6] presents a study on improving driver behavior in the sense of reducing fuel consumption, based on feedback from some data recorders. Studies have also been carried out to determine a customized route search method based on vehicle behavior recognition, [7]. Research has also been carried out on the influence of vehicle design and traffic simulation, on reducing fuel consumption, [8]. More articles on Eco-driving and eco-driving effectiveness have been published. Some experiments were carried out by integrating a driving simulator and a traffic simulator [9, 10, 11, 12].

This paper presents the results of a study on the use of a quality management tool - tree diagram, to solve the problem of reducing fuel consumption in vehicles in operation.

Quality management tools are techniques that apply to different areas of activity. Most techniques are graphical methods, which, in their clear and explicit form, form a graphic image suggestive to solve a problem or a nonconformity. Applications of quality management tools are found in several studies and research on solving some quality problems. The authors of this paper (unique authors or research teams) have published other studies on the use of fishbone diagram and tree diagram [13, 14, 15, 16, 17, 18, 19]. In the Specialty literature are studies which use the tree diagram to solve some quality problems in various fields, such as: machine tool manufacturing [20,21], the design of the services of medical assistance [22], waters quality [23], quality products according to customer needs [24], program management in different organizations [25], etc.

2. Classification of actions to reduce fuel consumption in road vehicles in service

We studied the main factors that determine the fuel consumption of road vehicles in operation: the purchased vehicle model, driving style and maintenance. This has identified the actions needed to reduce fuel consumption. The actions have been grouped into 3 corresponding categories for the following three specific objectives: the acquisition of an automobile appropriate to the needs; the proper use of the vehicle (eco-driving); proper maintenance.

The study was based on a Brainstorming session with specialists in the field of motor vehicle engineering. Several papers in the literature were also consulted. The study was completed by drawing up a list of actions, presented in Table 1.

Table1. Actions to reduce fuel consumption

Specific Objective	**Actions**
1. Acquisition of a vehicle suitable for the needs	Avoiding the option for the automatic gearbox;
	For city journeys, "4 x 4" vehicles are not recommended;
	Choosing aluminum rims that reduce the weight of the car and contribute to fuel economy;
	Rationalization of facilities offered by electrical accessories;
	Choosing a diesel engine is recommended at a

	yearly run higher than the usual one (10 ÷ 15 thousand km / year).
2. Appropriate use of the vehicle (eco-driving)	Choosing shorter and less crowded routes;
	Choosing good quality roads;
	Adopt a preventative and continuous driving style with avoiding brakes / accelerators and unnecessary speed changes (no starting, accelerating and sudden braking);
	Avoiding left-right maneuvers through the strips, turns standing on the spot;
	Using the motor brake as much as possible (speed reduction to be done gradually)
	No long stops (more than one minute) with the engine running;
	Avoiding the initial engine heating for too long;
	Observing speed limits and moderate speed travel;
	Rational use of air conditioning only when strictly necessary;
	Using a GPS navigation system (the system offers information to the driver and helps in vehicle fleet management at auto monitoring);
	Do not place unnecessary luggage (especially heavy) in the vehicle;
	Periodic checking of the tire pressure (suitably inflated tires);
	Do not use in the summer, winter tires that have increased consumption;
	Using proper fuel;
	Using the control system of cruising speed.
3.Proper maintenance	Maintaining an optimal state of operation of the car through regular maintenance by the specialist;
	Performing the periodic review according to the manufacturer's recommendations;
	Comply with the instructions in the vehicle's manual on changing the air filter, spark plugs, engine oil, other fluids, etc .;
	Dealing with deficiencies is done by professionals;
	Checking and adjusting the engine is done by the professional mechanic.

3. Tree Diagram for reducing fuel consumption

A tree diagram is proposed on 3 levels: the general objective, three specific objectives and the actions corresponding to each specific objective. The diagram is shown in Figure 1.

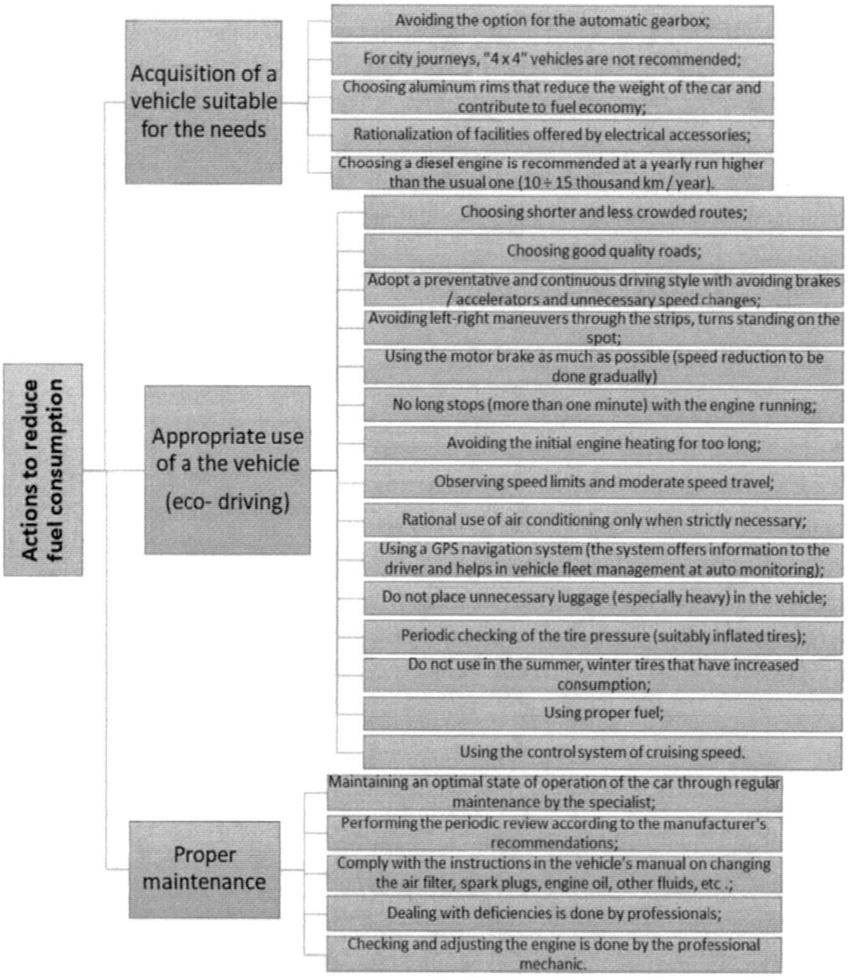

Figure 1. Tree Diagram of reducing fuel consumption

The general objective forms the level I and is given by the problem to be analyzed: reducing fuel consumption. The second level of the diagram is given by the three specific objectives: the acquisition of a vehicle suitable for the needs; proper use of the vehicle (eco-driving); proper maintenance. The actions proposed to solve the problem and the fulfillment of the three specific objectives are the third level of the tree diagram

4. Conclusion

The tree diagram presented in this paper is used to answer the question: How can fuel consumption be reduced? The chart provides the optimal solutions to solve a major problem that concerns all of the world, reducing fuel consumption.

The specified tree diagram can be used in service and showroom of vehicles. The diagram can be displayed as a poster, at a visible plae, to be read by drivers who go to buy a car or to perform service operations. The diagram provides a complete and suggestive picture of the actions to be taken to reduce fuel consumption. Some economic driving tips are needed and the probability of being applied increases because the solutions come from some specialists in the automotive industry.

The authors of the paper believe that the tree diagram obtained is a good tool for informing and raising the driver's awareness of his role in reducing fuel consumption.

Acknowledgments
This work received funding from the Technical University of Varna.

References

[1] Al-Rousan A A 2010 Reduction of fuel consumption in gasoline engines by introducing HHO gas into intake manifold *Fuel and Energy Abstracts* **35(23)** pp 12930-35.

[2] Soofastaei A, Aminossadati S, Kizil M S and Knights P 2016 Reducing Fuel Consumption of Haul Trucks in Surface Mines Using Artificial Intelligence Models, *Conference: Coal Operators' Conference* (At University of Wollongong) Vol. 1.

[3] Lo C , Chen C , Kuan T , Lo K and Cho H, Fuel Consumption Estimation System and Method with Lower Cost 2017 *Symmetry* **9** p 105.

[4] Lee W C , Tsang K F, Chi H R, Hung F H, Wu C K, Chui K T, Lau W H and Leung Y W 2015A high fuel consumption efficiency management scheme for PHEVs using an adaptive genetic algorithm *Sensors* **15** pp 1245–51.

[5] Haworth N and Symmons M Driving to reduce fuel consumption and improve road. http://acrs.org.au/files/arsrpe/RS010036.pdf.

[6] Toledo G and Shiftan Y 2016 Can feedback from in-vehicle data recorders improve driver behavior and reduce fuel consumption *Transp. Res. Part A Policy Pract.* **94** pp 194–204.

[7] Bao Y and Chen W 2016 A personalized route search method based on joint driving and vehicular behavior recognition *Proc. of the IEEE MTT-S International Wireless Symposium*, (Shanghai, China).

[8] Lin Y, Chen P and Chan K 2017 A method for reducing fuel consumption of urban scooters using vehicle design and traffic simulation, *Proc.of the Institution of Mechanical Engineers, Part C: Journal of Mechanical Engineering Science*, Vol.231 issue: 7 pp: 1252-71

[9] Matsumoto S , Park T and Kawashima H 2014 A Comparative Study on Fuel Consumption Reduction Effects of Eco-Driving Instructions Strategies *International Journal of Intelligent Transportation Systems Research* **12-1** pp 1–8.

[10] Saito A and Takada H 2008 The relationship between eco-driving patterns of light duty freight vehicles and the improvement of fuel economy in real traffic conditions. *Proc. of JSAE before lecture presentation* 99–08 pp 27–30.

[11] Hiraoka T, Terakado Y, Matsumoto S and Yamabe S 2008 Quantitative evaluation of eco-driving on fuel consumption based on driving simulator experiments *Proce. of 7th World Congress on Intelligent Transport Systems* pp 163–68.

[12] Hiraoka T, Terakado Y, Matsumoto S and Yamabe S 2011 Effect of eco-driving instruction and fuel consumption meter on fuel consumption reduction, *Hum. Interf. Soc. J.*. **12**(1) pp 71–78

[13] Luca L and Filip CP 2011 On the assessment of the public order services quality by using classic instruments of quality management, *Proc. Int. Conf. The Knowledge- Based Organization* (Sibiu) pp 691-6.

[14] Luca L 2015 The Study of Applying a Quality Management Tool for Solving Non-Conformities in an Automotive, *Applied Mechanics and Materials Trans Tech Publications*, vol 809-810 pp 1257-62.

[15] Luca L, Todorut A V and Luca T O 2018 Quality management applied to analyze the reduction of the pollution which is generated from road transportation in aglomerated urban areas *Scientific Bulletin of Naval Academy (SBNA)* vol. XXI, Issue no.1, (MBNA Publishing

House Constanta) pp 116-22.

[16] Cirtina L M, Cirtina D and Luca 2014 Quality Management in Projects - Quality Planning, *Applied Mechanics and Materials* **657** pp 891-5.

[17] Luca L 2016 A study on quality analysis measuring process *Fiability & Durability* **2** pp 68-72.

[18] Luca L , Pasare M and Stancioiu A 2017 Study to determine a new model of the Ishikawa diagram for quality improvement *Fiability & Durability* **1** pp 249-54.

[19] Pasare M 2016 On the mitigation actions of defects for plastical deformed parts *Fiability & Durability* **2** pp 114-7.

[20] Ghimisi S and Nicula D 2018 Actions to increase the reliability of chain transmissions *22nd International Conference on Innovative Manufacturing Engineering and Energy - IManE&E, Issue MATEC Web Conf.* Vol.178, article number 06002 DOI: https://doi.org/10.1051/matecconf/201817806002

[21] Ghimisi S 2018 Measures for correct design, manufacturing and exploitation of gear transmissions to increase the reliability, *Scientific Bulletin of Naval Academy* Vol.21 (Constanta) pp 103-8.

[22] Despotou G, Jones RW *a*nd Arvanitis TN 2016 *Using Event Trees to Inform Quantitative Analysis of Healthcare Services Unifying the applications and foundations of biomedical and health informatics. Book Series: Studies in Health Technology and Informatics* Vol.226 pp 119-122.

[23] Saghebian S M, Sattari M T, Mirabbasi R and Pal M 2014 Ground water quality classification by decision tree method in Ardebil region, Iran. *Arabian Journal of Geosciences* **7 – 11** pp 4767-77.

[24] Islam M N 2004 A methodology for extracting dimensional requirements for a product from customer needs *Int.J.of Advanced Manufacturing Technology* **23 Issue: 7-8** pp 489-94.

[25] Li SS , Lee LC 2011 Using fishbone analysis to improve the quality of proposals for science and technology programs *Research Evaluation* **20-4** pp 275-82.

[26] Ionita I 2008 *Managementul calitatii si ingineria valorii.* (Editura ASE Bucuresti).

IOP Conf. Series: Materials Science and Engineering **614** (2019) 012006 doi:10.1088/1757-899X/614/1/012006

An expert system for vehicle accident reconstruction

D Lyubenov*, V Mateev and G Kadikyanov

University of Ruse, 8 Studentska Str., 7017 Ruse, Bulgaria

*E-mail: dlyubenov@uni-ruse.bg

Abstract. The paper presents the capabilities of the developed expert system for vehicle accident reconstruction. The system determines converted energy for each vehicle, post-collision and pre-collision vehicles speeds, vehicle braking deceleration, the times in the braking process and total stopping distance. This system provides an opportunity to investigate the impact of various parameters at vehicle accident reconstruction. The developed expert system significantly reduces the time for the preparation of the expertise and the admission of errors. It can be used in expert practice and training.

1. Introduction

Bulgaria is one of the countries in the European Union with the highest number of deaths in road accidents, according to the population [1].

Under the Law on the Judiciary to investigate the causes and the possibilities for prevention of traffic accidents expertise are appointed for the needs of the court process. It is impossible to have quality justice without quality expertise. This is because many of the cases are dealt with primarily on the basis of the expert reports. Poor expertise leads to poor justice.

In Bulgaria for each judicial district compile lists of specialists approved experts. The procedure for inclusion in these lists of specialists qualified as experts is regulated by the Ministry of Justice.

Unfortunately lists of experts shall be approved by the Commission which includes only judges and prosecutors – persons who do not have a technical education. These Commissions have no knowledge and cannot assess the technical competence and ability of the candidates for experts. Moreover, there is no requirement for licensing experts in Bulgarian legislation. All of this can lead to the production of low-quality expertise.

In order to increase the quality of the expertise, the Rousse University team has developed a vehicle accident reconstruction expert system based on the energy method. This article presents the possibilities and the way this expert system works. There are no suitable software applications in Bulgaria that can be used by the experts, which is a further reason for doing so.

2. Mathematical model

According to [2, 3, 4, 5, 6] one of the most important issues in the investigation and analysis of road accidents is the determination of vehicle speeds and the stopping distance.

In practice, experts use different methods [2, 7, 8, 9, 10, 11, 12, 13, 14, 15] for reconstruction of vehicle collision accidents. Each has its advantages and disadvantages. In some cases, one method is better, and in others - another.

In Bulgaria, one of the commonly used method for determining the vehicles speed in accidents is the energy method. This method is to use six points damage profile depth to determine the dimensions of the crush [16, 17, 18, 19, 20]. Under this method, the converted energy E for each vehicle during the impact is calculated as follows:

$$E = (1+tg^2\alpha)\frac{W}{5}\begin{bmatrix} \frac{A}{2}(d_1+2d_2+2d_3+2d_4+2d_5+d_6)+ \\ +\frac{B}{6}(d_1^2+2d_2^2+2d_3^2+2d_4^2+2d_5^2+d_6^2+ \\ +d_1d_2+d_2d_3+d_3d_4+d_4d_5+d_5d_6)+2,5A^2/B \end{bmatrix}, J \quad (1)$$

where α is angle between the normal to the damage profile surface and impact pulse vector; W - damage profile width, m; $d_{1\text{-}6}$ - damage profile depth, cm; A and B - crash coefficients, N/cm and N/cm^2.

The change in vehicle velocity ΔV can be written as in equation:

$$\Delta V_{1,2} = \left[\frac{2m_{2,1}(E_1+E_2)}{m_{1,2}(m_1+m_2)}\right]^{1/2}, m/s \quad (2)$$

where m_1 is mass of vehicle 1; m_2 - mass of vehicle 2; E_1 - converted energy of vehicle 1; E_2 - converted energy of vehicle 2.

The post-collision velocities u can be written as in equation:

$$u_{1,2} = \left[2g(\varphi_{1-n}S_{1-n} + \varphi R\omega) \pm V_{l(a)}{}^2\right]^{1/2}, m/s \quad (3)$$

where $\varphi_{1\text{-}n}$ is coefficients of friction; $S_{1\text{-}n}$ - distance travelled by the vehicle after the collision; g - gravity acceleration; R - radius of vehicle rotation, m; ω - rotation of the vehicle after the collision, rad; $V_{l(a)}$ - loss (acquired) speed of vehicles after impact.

The radius of vehicle rotation can be calculate by the following equation:

$$R = \left[\left(\frac{L_b}{2}\right)^2 + \left(\frac{B_\kappa}{2}\right)^2\right]^{1/2}, m \quad (4)$$

where L_b is vehicle wheelbase; B_k - vehicle track.

The pre-collision velocity can be written as in equation

$$V_{1,2} = \left(\Delta V_{1,2}^2 - u_{1,2}^2 \sin^2\theta_{1,2}\right)^{1/2} + u_{1,2}\cos\theta_{1,2}, m/s \quad (5)$$

where $\theta_{1,2}$ is angles between the velocities before and after impact.

To analyze accidents and determine the possibility of their prevention by drivers it is necessary to calculate the stopping distance. Total stopping distance can be calculate by the following equation:

$$D = tV + \frac{V^2}{2a}, m \quad (6)$$

where t is total start time for braking, s; V - vehicle speed, m/s; a - braking deceleration, m/s^2.

The braking deceleration for a longitudinal slope of road is calculated as follows

$$a = \left(\frac{\varphi\cos\beta}{c_e} \pm \sin\beta\right) g, m/s^2 \quad (7)$$

where β is slope of road angle, deg. c_e - coefficient of efficiency.
"plus" sign - slope up; "minus" sign - slope down.

The braking deceleration for a horizontal road is calculated as follows

$$a = \frac{\varphi g}{c_e}, m/s^2 \quad (8)$$

The total start time for braking is calculated as follows

$$t = t_d + t_v + 0,5t_i, s \tag{9}$$

where t_d is driver reaction time, s; t_v - vehicle reaction time, s; t_i - time for increase the braking deceleration, s.

3. Expert System

In the field of transport and traffic safety various calculation tools are used. Unfortunately they have a number of limitations, such as: there is no possibility of entering specific values for different parameters; do not provide the ability to select pre-set reference data; do not inform the user about incorrect data entered etc. [11, 21 - 24]. Because of these features, they are not suitable in expert practice.

The main window of the energy method based expert system for vehicle accident reconstruction is presented in figure 1.

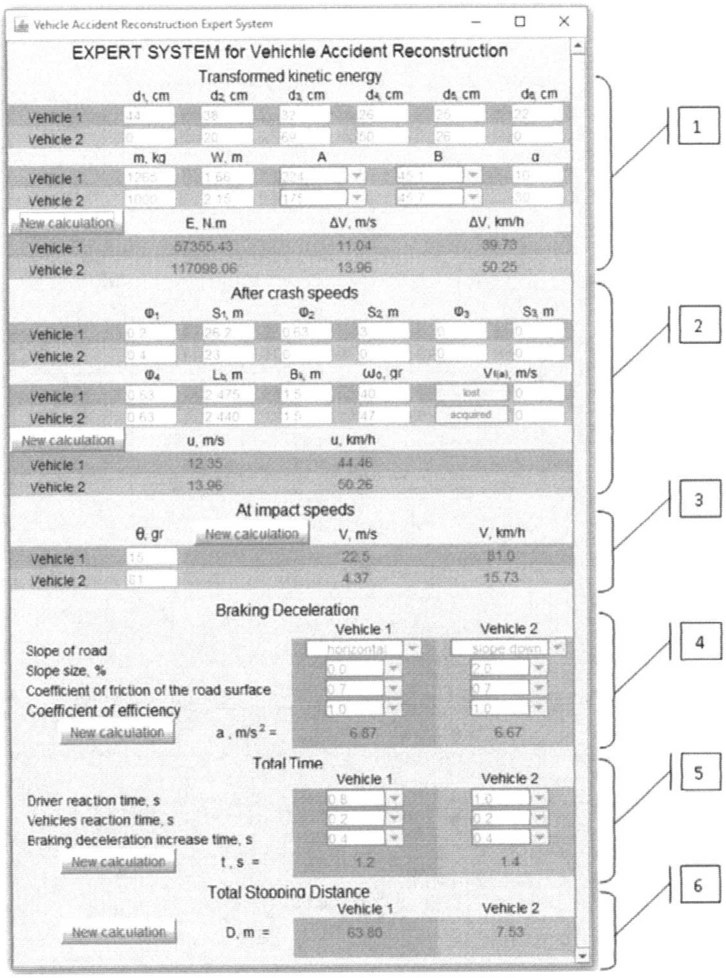

Figure 1. Main window of expert system for Vehicle Accident Reconstruction.

The expert system is developed by means of the object-oriented and platform independent programming language – Java [25]. It makes possible for the application to run on every computer platform that has Java Virtual Machine installed on it both as a standalone application and as an applet integrated into a web page and available on the Internet.

The expert system consists of six panels in a scrollable window. The first one (figure 1, position 1) is for calculation of the transformed kinetic energy (1) and the change in vehicles speeds at impact (2), the second (figure 1, position 2) – for the post-collision velocities (3) and the third (figure 1, 3 positions) – for the pre-collision velocities (4). The fourth panel is for determination of the braking deceleration, the fifth – for determination of the total time and last sixth panel is for the determination of the total stopping distance.

3.1. Functioning of the Expert System

After starting the system the user fills in panel 1 the measured values for damage profile depths in fields d_1 - d_6, the masses of vehicles m, damage profile width *W*, crash coefficients *A* and *B*, the angle α between the normal to the damage profile surface and impact pulse vector and clicks the "Calculate" button. If it turns out that in some fields are entered non-numeric values a warning message appears stating that the value of the specified field is not numerical (figure 2, position 1).

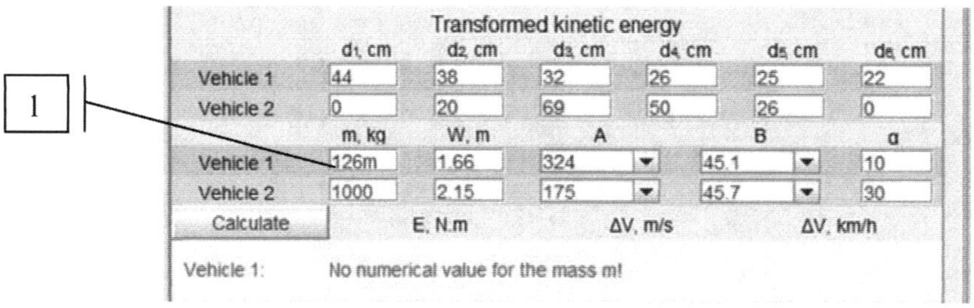

Figure 2. Warning message for a not numerical value.

After checking for not numerical values, starting from the left margin of the top row and ending with the right margin of the bottom row another checking is performed that determines if any value falls within the pre-set space of values for it or is equal to a value of a number of acceptable values. If some value is unacceptable (figure 3, position 1) a message about it appears.

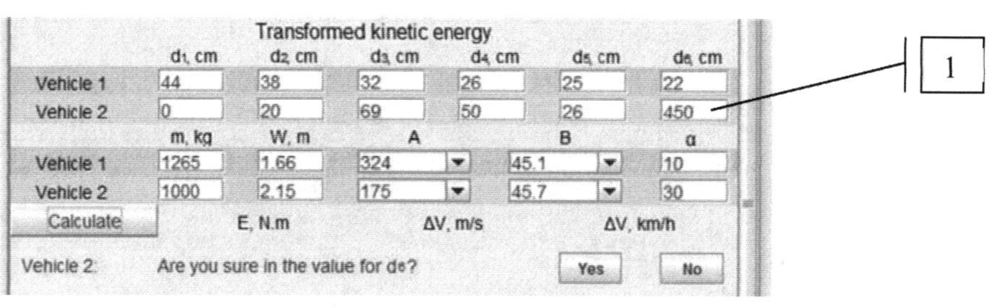

Figure 3. Warning message for unacceptable value.

If the user clicks the "No" button the entered in the field value is deleted (figure 4, position 1) and a new message appears (figure 4, position 2).

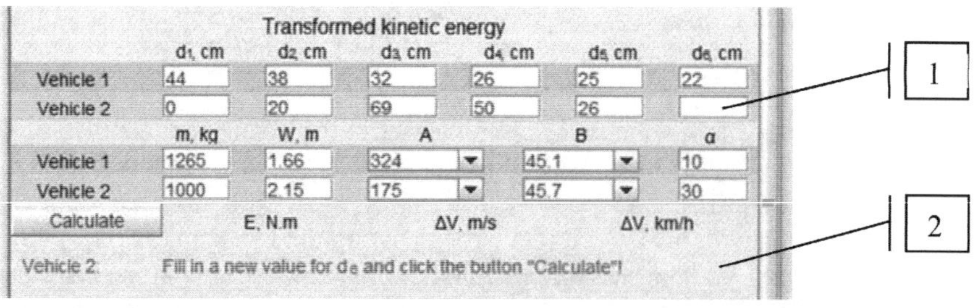

Figure 4. Warning message for a new value.

If the user presses the "Yes" button the entered in the field value is reserved and the message "The value of d_6 is accepted. Press "Calculate" to continue". So the entered value, though not belonging to a preset space of values for it, is accepted and it will be used in the upcoming calculations.

In order to determine the speeds of the vehicle after the collision in panel 2 (figure 1) it is necessary to fill in coefficients of friction φ and corresponding movements of the vehicles S, wheelbase L_b and rear track B_k of the vehicles and rotation of the vehicles after the collision ω_o.

There is a possibility to determine the speed when driving on three surfaces with different coefficients of adhesion. The impact of the lost or acquired velocity of the vehicles after the collision is taken into account (toggle button lost/acquired). Described above checks for incorrect or non-numerical values are made in this panel as well.

At impact speeds of the vehicles are calculated in panel 3. For this purpose it is necessary to fill in the angles between the velocities of the vehicles before and after the impact θ (figure 1). The other values, necessary for determining the speeds, are automatically generated from the calculations in the previous two panels of the system.

The fourth panel provides controls for determining the braking deceleration, indicating the presence and magnitude of road slope as well as the coefficients of friction and efficiency. To determine the presence and type of road incline, a combo box with three items is used: horizontal, slope up or slope down (figure 5, position 1). In the slope size field, a value is selected or entered, and this is only possible for a slope road (figure 5, position 2). Sometimes, in expert practice, the friction and efficiency coefficients must be determined experimentally by different methods and tools. The corresponding values are set in editable combo boxes.

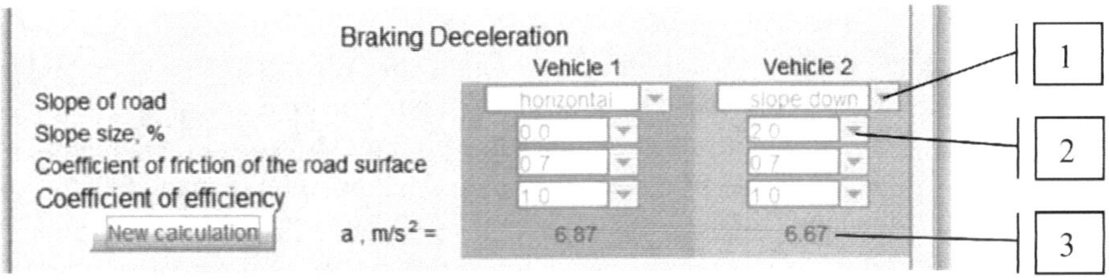

Figure 5. A general view of the fourth panel.

When the user clicks the "Calculate" button, the braking deceleration value is calculated and displayed in a non-editable text field (figure 5, position 3). Then the controls for setting the values of the values with which the braking deceleration is calculated become inactive. To calculate the braking deceleration with other values the user clicks the "New calculation" button, where the braking deceleration value is replaced by a dash (-), and the "Calculate" button and the controls become active.

To determine the total time, the controls on the fifth panel are used. They are editable combo boxes. The user can choose from a set of predefined values or can input her/his own values for driver and vehicle reacting times and the time for increase the braking deceleration. The total time value is calculated and displayed after pressing the Calculate button. Then the fields become non-editable, and the button changes to "New Calculation".

In the last sixth panel, there is only the "Calculate" button and two text fields that display the values for the stopping distances for both vehicles.

Panels 4, 5 and 6 also provide a check for incorrectly entered or missing values as well as for values beyond the predefined limits. For each established inaccuracy, the expert system shall display the appropriate alert message. This significantly reduces the possibility of making mistakes.

3.2. Realization of the described functionality

For each panel there is an array for correctness of entered data. Each element of this array corresponds to the value entered in a particular text field. At the beginning all elements of the array are initialized to 0, which corresponds to the incorrect data. At every click of the "Calculate" button its elements are reset. Then the data in each field is evaluated and if it's value is numerical, to the corresponding element in the array is assigned 1. Then the elements of the array are checked from the beginning to the end and at first meeting of an element showing a non-numerical value, i.e. 0, a message that in the relevant field is introduced non-numerical value appears. After entering a new value these tasks are performed again. This continues until in all text fields there are only numerical values. Then entered in each field value is checked for belonging to a predetermined range of acceptable values. When the checking comes to a value that is not in the preset interval a panel with a message for confirmation of the value appears. If the value is accepted it is stored in an array for accepted values and is noted in an array for acceptable values. It does not matter that the value is not in the predefined interval or set of values. So at the end in all fields there are acceptable values or values, that are accepted explicitly for correct.

To determine the content of the warning dialogues, a one-dimensional array with 3 elements is used. The first element corresponds to the value for the braking deceleration, the second – to the total time and the third - to the pre-collision velocity. Each element can be equal to 1 if the corresponding value is determined and 0 if it is not. When the values of all three elements are equals to 1, the stopping distance can be calculated. Otherwise, a warning message appears indicating which value has not yet been determined.

The developed expert system for determining speeds at impact between vehicles can be used for a strike of a vehicle in a solid object (tree, pillar, wall, etc.) as well. The system reduces significantly the time for determination of the speeds and allows to determine the influence of various parameters on final results. The system prevents errors in the calculation, thus contributing to improving the quality of expert report.

4. Conclusions

The expert system for vehicle accident reconstruction has been developed. The expert system is developed by means of the object-oriented and platform independent programming language – Java. The possibilities and the way of operating are presented. The expert system consists of six panels in a scrollable window. The first one is for calculation of the transformed kinetic energy and the change in vehicles speeds at impact, the second – for the post-collision velocities, the third – for the pre-collision velocities, the fourth panel - for determination of the braking deceleration, the fifth – for determination of the total time and the last sixth panel is for the determination of the total stopping distance.

The system significantly increases the efficiency of experts' work, reduces the probability of making mistakes and provides opportunity to investigate the impact of various parameters at vehicle accident reconstruction. It can find a real application in the vehicle accident reconstruction and training of experts.

5. References

[1] European Commission. Transport. Statistical pocketbook 2017. Available at: https://ec.europa.eu/transport/facts-fundings/statistics/pocketbook-2017_en.
[2] Lang W, Biao G and Tao C 2013Vehicle Continuous Collision Accident Reconstruction System Development *Procedia - Social and Behavioral Sciences* vol 96 pp 1659-1669.
[3] Han I and Kang H 2016 Determination of the collision speed of a vehicle from evaluation of the crush volume using photographs J. *Automobile Engineering* vol 230(4) pp 479-490.
[4] Lozana J A, Vera C and Félez J 1998 A computational dynamical model for traffic accident reconstruction *Int J Vehicle Design* vol 19 (2) pp 213-227.
[5] Park S and Han I 2010 Using frontal crush for collision speed estimation in reconstruction analysis of uncertain traffic accidents *J Transp Res* vol 17(2) pp 49–60.
[6] Zhang X Y, Jin X L and Qi W G 2006 Analysis and reconstruction of the typical traffic accident based on the tire marks *J Basic Sci Eng* vol 14 (3) pp 418-426.
[7] Bartlett W and Fonda A 2003 Evaluating uncertainty in accident reconstruction with finite Differences *SAE Technical Paper* No. 2003-01-0489 Warrendale PA.
[8] Brach R M and Matthew R B 2011 Vehicle Accident Analysis and Reconstruction Methods *SAE International* Second Edition p. 418.

[9] Evtiukov S, Kurakina E, Lukinskiy V and Ushakov 2017 A Methods of Accident Reconstruction and Investigation Given the Parameters of Vehicle Condition and Road Environment *Transportation Research Procedia* vol 20 pp 185-192.
[10] Han I and Kang H 2016 Three-dimensional crush scanning methods for reconstruction of vehicle collision accidents *International Journal of Automotive Technology* vol 17(1) pp 91-98.
[11] Kumella A 2016 Building Rail-Vehicle Braking Distance Calculation Tool Using C# Programming Language *International Journal of Scientific and Research Publications* vol 6 iss 7 pp 162-167.
[12] Lang W and Tao C 2003 Study of the three-dimensional vehicle impact accident simulation and reconstruction system *China Journal of Highway and Transport* vol 16(4) pp 83-86.
[13] Martínez F, Páez H, Furones A and Sánchez S 2016 Pedestrian-Vehicle Accidents Reconstruction with PC-Crash®: Sensibility Analysis of Factors Variation *Transportation Research Procedia* Vol. 18 pp 115-12.1
[14] Wach W 2016 Calculation reliability in vehicle accident reconstruction *Forensic Science International Journal* vol 263 pp 27-38.
[15] Wach W and Unarski J 2007 Uncertainty of calculation results in vehicle collision analysis *Forensic Science International Journal* vol 167 (2) pp 181–188.
[16] Franck H, Franck D 2009 Mathematical Methods for Accident Reconstruction *CRC Press - Taylor & Francis Group* p. 303.
[17] Gabauer D J and Gabler H C 2006 Comparison of delta-V and occupant impact velocity crash severity metrics using event data recorders *Annual Proceedings/Association for the Advancement of Automotive Medicine* vol 50 pp 57–71.
[18] Johnson N S and Hampton C G 2014 Evaluation of NASS-CDS Side Crash Delta-V Estimates Using Event Data Recorders *Traffic Injury Prevention Journal* vol 15 iss 8 pp 827-834.
[19] Pride R, Giddings D, Richens D and McNally D 2013 The sensitivity of the calculation of ΔV to vehicle and impact parameters *Accident Analysis & Prevention* vol 55 pp 144-153.
[20] Shelby S G 2011 Delta-V as a Measure of Traffic Conflict Severity *3rd international conference on road safety and simulation* Indianapolis USA.
[21] Calculate stopping and braking distance Available at: http://www.countcalculate.com/cars-and-speed/stopping-braking-distance.
[22] Calculator for Stopping (Braking) Distance, Acceleration/Deceleration, Speed, Time Available at: https://www.johannes-strommer.com/diverses/pages-in-english/stopping-distance-acceleration-speed/.
[23] Road Safety Week. Available at: http://www.roadsafetyweek.org.uk/stopping-dist-calc/story_html5.html.
[24] Vehicle Stopping Distance Calculator. Available at: http://www.csgnetwork.com/stopdistcalc.html.
[25] Simon R, Heller P and Ernest M 2001 Complete JavaTM 2 Certification *Study Guide SoftPress Ltd* p 780.

Acknowledgments
The study was supported by project of University of Ruse “Angel Kanchev”, № 19-ФТ-01, “A study of advanced technology to reduce automotive emissions”.

IOP Conf. Series: Materials Science and Engineering **614** (2019) 012007 doi:10.1088/1757-899X/614/1/012007

Research on the Possibility of Increasing the Power Output of a High - Speed Diesel Engine by Adding Compressed Natural Gas

K Bogdanov, R Dimitrov *

Technical University of Varna, Department "Transport Engineering and Technologies", 9010 Varna, Bulgaria

* E-mail: r_dimitrov@tu-varna.bg

Abstract: The paper presents studies on a high-speed diesel engine driven by added CNG (compressed natural gas) in order to increase its power. The effects of the amount of CNG added on the engine performance have been studied. Throughout the tests the effective performance and pollution emission indicators of ICE (Internal Combustion Engine), the cylinder pressure, etc. were measured. Under all operating conditions the diesel fuel consumption per hour and the rotation speed have remained constant. The amount of CNG added is variable. As a result of the research, relations have been established concerning the overall efficiency, the air to fuel ratio for steady-state performance and smoke opacity. An increase in CO, NOx and HC emissions has been recorded and at the same time the amount of CO_2 released has decreased. Both the maximum pressure in the combustion chamber and the pressure rise rate increase which results in rising the engine performance harshness. When CNG has been added in quantity about 47% of the total amount of fuel, the peak value is achieved. The influence of the amount of added gas fuel on the calculated gross heat release rate has been determined. The maximum heat release rate does not change significantly. The nature of the combustion process does not change either. In conclusion, findings validate the possibility to achieve up to 30 % higher maximum rated power by adding 56% CNG mass fraction of the total amount of fuel. Provided the amount of CNG is up to 48% of the total amount of fuel, the ICE ecological indicators have not been deteriorated, furthermore the power has been increased by up to 20%.

1. Introduction

The continuous increase in the specific volume of the modern piston ICE is the result of the investigations conducted by researchers. These investigations are focused on improving engine energy efficiency.

A well-known fact is that diesel engines do not utilize the air introduced to the cylinders efficiently, compared to engines operating with homogeneous fuel-air mixtures [1]. This disadvantage is overcome either by cylinders overfilling or by finding alternative combustion processes.

Another alternative method is adding hydrocarbon gases such as propane-butane, methane and others during the diesel engine operation. Thus a certain homogenization of the mixture has been achieved furthermore there exists the possibility of lowering the smoke opacity limit at air to fuel ratio of $\lambda = 1.05 \div 1.1$.

This method is mostly used in heavy-duty truck engines, the fleet transportation industry and stationary diesel engines, running in modes close to their full load [2]. On the one hand, from the economics point of view, the engine performance indicators are improved, while on the other hand the CNG method promotes environmental friendliness with lowering the quantities of soot deposited. There are certain advantages of adding gas fuel to the diesel engine: the engine power is increased since more air is being introduced to the cylinder without exceeding the smoke opacity limit.

2. Experimental methodology

A high frequency direct injection diesel engine Prima 65, produced by Perkins Engines Ltd" England is used for the tests. The main technical data are as follows: rated power of 46kW/4500 min-1; the maximum engine torque of 122N.m/2500 min^{-1} and engine capacity of 1993cm^3 .

Throughout the tests the following characteristics were measured: engine load; crankshaft rotation speed; diesel engine air consumption; fuel and CNG consumption; exhaust gas temperature; coolant temperature; lube oil temperature; ambient temperature; oil pressure in the lubricating oil system; barometric pressure; air consumption; smoke opacity and ratio of CO; CO2; O2; HC; NOx emissions in the exhaust gases; injector needle movement and mean effective pressure in the combustion chamber. In all modes the diesel fuel consumption per hour remained constant– 4,0 l/h. The quantity of CNG added varied.

The test was carried out at a constant frequency mode of 2500 min^{-1}. Large black markers on the graphs show in comparison the parameter values for both diesel fuel and maximum torque of 122Nm. They are positioned close to the marker at which the same mean effective pressure with CNG added has been achieved.

3. Experimental data and results

By adding 56% of the total amount of CNG, a torque of 153,8Nm and the respective power of 43,3kW were achieved. The torque obtained was 26% bigger than the original one (122N.m) according to the production characteristics of the engine. This limitation was imposed by the smoke opacity ratio, permissible for this type of engine - 55%. Taking into account that the power output was close to the rated power, and the rotational speed was 56% of the rated one, it should be assumed that the absolute amount of soot deposited in the atmosphere led to the same ratio. When the smoke opacity was equal to the smoke opacity caused by diesel fuel (D=33%) only, an increase in the torque was observed by 20%(Fig.1).

The following most important dependencies have been outlined by the research conducted:

- The engine effective efficiency (η_e) is close to the efficiency in the cases when the engine runs only on diesel fuel. A steady-state performance of the engine is obtained with air to fuel ratio of λ= 1,10 and smoke opacity D=55%. A certain increase in the percentage ratio of the CO and HC emissions is observed. The quantity of the carbon dioxide (CO_2) released is less under similar operating conditions and same power due to the change of the fuel composition. The nitrogen oxides are at higher levels and eventually they are reduced to the original specification levels due to the lower concentration of free oxygen in the cylinder during the combustion process, caused by lowering of the fuel to air ratio.
- Indicator parameters show that with the increase in the percentage of CNG and load respectively, the engine experiences an increase in harshness as well. The maximum pressure in the combustion chamber (p_{max}) increases and reaches its maximum point when 47% CNG of the total amount of fuel is added. The maximum rate of pressure increase ($dp/d\varphi_{max}$) is raised considerably (up to 1,93MPa/deg) which in turn causes an increase in the noise levels under operating conditions. The maximum cylinder temperature with a maximum of 56% CNG added is in linear relationship and is higher by about 20°K of the temperature under maximum load conditions.

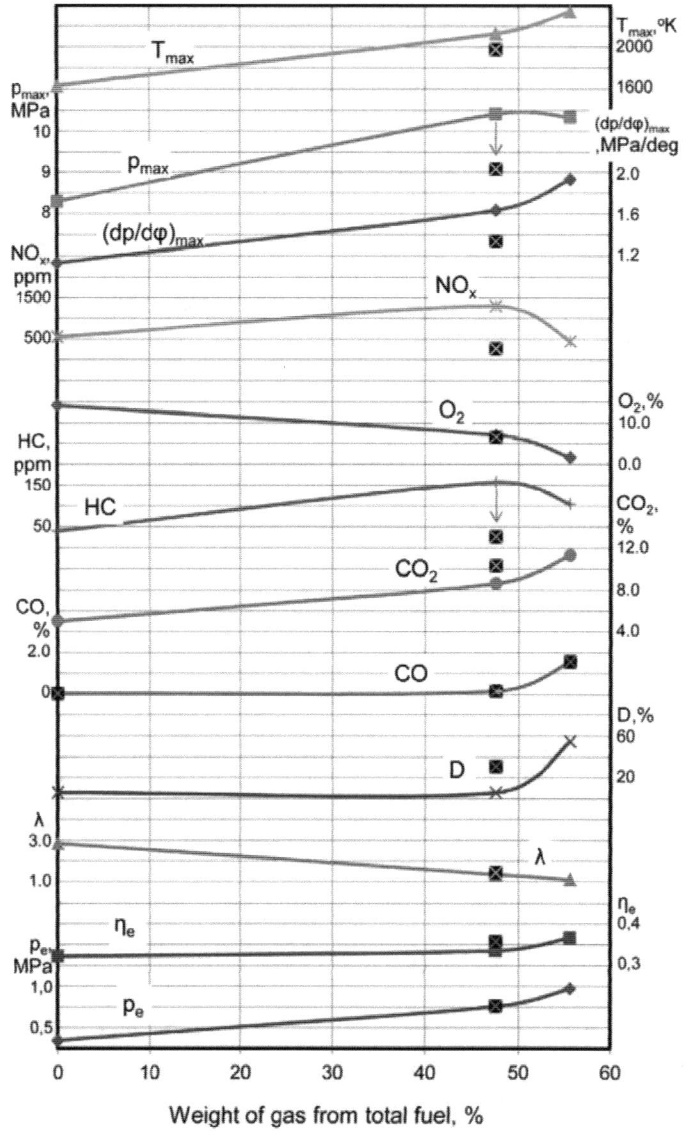

Figure 1. Influence of the part of gaseous fuel on the parameters of the engine
p_e – mean effective pressure, MPa; ηe – overall efficient; λ – Relative air/fuel ratio; D - Opacity of exhaust gas, %; CO - carbon monoxide of exhaust gas, %; CO_2 - carbon dioxide of exhaust gas, %; HC – hydrocarbons of exhaust gas, ppm O_2 - oxygen of exhaust gas, %; NO_x-nitrogen oxides of exhaust gas, ppm; (dp/dϕ)max - Rate of pressure increase, MPa/deg; p_{max} - Maximal cylinder pressure, MPa; T_{max} – Maximal temperature of burned gas, °K

Figure 2 shows the influence of the quantity of gas fuel added on the heat release rate. It is observed that the maximum heat release rate does not change significantly when the quantity of 48% CNG of the total amount of fuel is added. The heat release rate is higher in the case when 56%CNG of the total amount of fuel is added than when the engine runs only on diesel fuel. The beginning of the visible combustion does not alter significantly (only within 1° of the crankshaft). The end of the process of visible combustion does not change considerably either. The nature of the combustion process does not change, only the instantaneous maximum combustion rates increase.

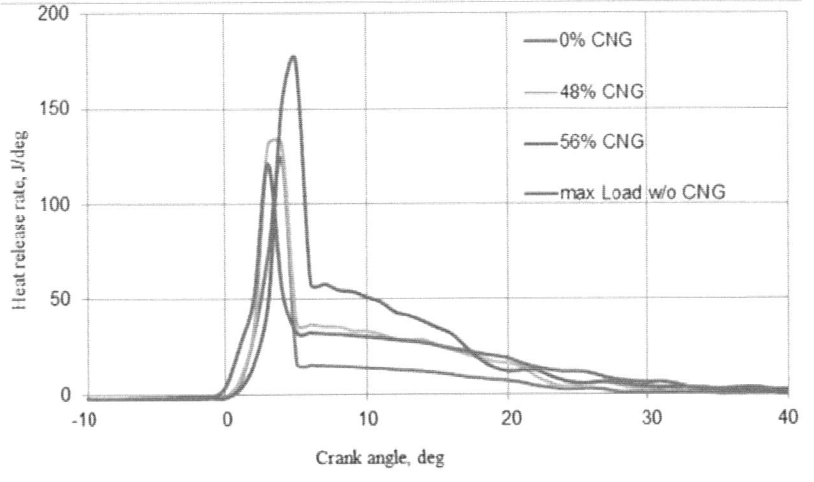

Figure 2. Influence of the quantity of gas fuel added on the heat release rate.

4. Conclusion

The following conclusions can be drawn from the present study:

- It is possible to achieve almost 30% of higher power coupled with economics characteristics by adding up to 56 % CNG of the total amount of fuel, complying with the smoke opacity limit permissible;
- Adding up to 48% CNG of the overall amount of fuel used, does not deteriorate the ICE ecological indicators, furthermore the power has been increased by up to 20%.
- Declining the amount of soot deposited in the atmosphere by up to 40% mass fraction at maximum load.

Acknowledgments

This work received funding from the Technical University of Varna by project NP12 with a project leader Assoc. prof. PhD eng. R. Dimitrov.

References

[1] Heywood J 1988 *Internal Combustion Engine Fundamentals* (McGraw Hill) p 372-375.
[2] Iliev S 2017 Investigation of n-butanol blending with gasoline using a 1-D engine model *3rd Int. Conf. Vehicle Technology and Intelligent Transport Systems* **1** 385–91.

IOP Conf. Series: Materials Science and Engineering **614** (2019) 012008 doi:10.1088/1757-899X/614/1/012008

Transient operation of a direct injection diesel engine

Z Ivanov*, V Mihaylov

Technical University of Varna, Department "Transport Engineering and Technologies", 9010 Varna, Bulgaria

* Corresponding author: zdravko.ivanov@tu-varna.bg

Abstract. A direct injection diesel engine was studied in transient operation. Experimentally is determined the injected fuel per cycle at varying initial and end load conditions, simulating operation at urban traffic. Phases have been identified in the change of parameters of fuel delivery and components in the exhaust gases, limited by emission standards. There is a potential for reducing the carbon emissions of the engine and the amount of emitted toxic components by influencing the duration of the transient processes.

1. Introduction

Automotive internal combustion engines operate under different conditions in terms of load and rotational speed. Studies show that most of the passenger mobility is carried out by private vehicles, with the average range of cars being driven in relatively short distances or in urban conditions [1]. The impact of cars in such conditions is characterized by predominant acceleration movements that are heavily dependent on the particular infrastructure and organization of the urban environment.

Operating modes of the engine in these cases are transient (non-stationary), associated with continuous change of speed and load, the processes being directly related to both road conditions and the need to switch gears. As a result of these changes in operating conditions, engine performance is significantly different from that of stationary operating modes.

Transient modes are characterized by the successive working processes in one engine cylinder under different thermodynamic conditions [2, 3]. As a result, there is an increase in fuel consumption and associated carbon emissions, as well as an increased content of the limited components in the exhaust gases [4]. In this sense, the actual operating conditions of the motor vehicles require constant updating of the tests provided for in the type approval procedures and the environmental standards [5].

The purpose of the present study is to examine the operation of the engine in transient modes and to determine the variation in fuel consumption and emissions of harmful components in exhaust gases. The main task of this report is to investigate the dependencies of changing cyclic quantity of fuel injected and the emissions of harmful components for a direct injection diesel engine.

The object of the study is the Perkins PRIMA 65 diesel engine. The main technical characteristics of the engine are: rated power $Pe = 46\ kW$ at crankshaft speed $n = 4500\ min^{-1}$; maximum torque $Me = 122\ N.m$ at rotational speed $n = 2500\ min^{-1}$; engine volume $Vh = 1.993\ dm^3$; cylinder diameter $Dc = 84.4\ mm$; stroke of the piston $Sh = 88.9\ mm$; number of cylinders - 4; compression ratio $\varepsilon = 18.1$; direct injection of fuel and water cooling. The engine is equipped with a fuel injection pump CAV-DPAM 3842 F020 and BDLL 150 S 6556 injectors.

2. Methodology

Experimental engine performance studies have been conducted. Engine operating modes are designed to match the most commonly used by cars when driving in urban areas. When driving in urban areas, engines work mostly on small and medium loads as well as a significant part of the time in idling mode. Rotational speeds are low and are usually in the area around and slightly above the maximum torque of the engine. The movement of the car is characterized by frequent stops and start off, as well as intensive gear changes. Increasing and decreasing of rotational speed are also accompanied by a corresponding increase and decrease in engine load. From operational and environmental point of view, the carbon emissions and the environmental performance of the engine in the positive acceleration modes are of interest.

The study was carried out on a specialized stand. Thermodynamic parameters and all variables necessary to determine the cyclic fuel quantity and the content of the harmful components in the exhaust gases are determined. The recording of measured variables is made with digital storage oscilloscope at high sampling rate. The workflow parameters for each of the working modes are determined directly or calculated. The in-cylinder pressure (indicator diagram), injection pressure, needle stroke, exhaust gas content are recorded.

The injected fuel per cycle is calculated. The injection characteristics (differential and integral) are computed by formulas *(1)* and *(2)*:

$$\frac{dq(\varphi)}{d\varphi} = \mu F(\varphi).\left[\frac{2}{\rho_g}.\left(p_g(\varphi) - p_i(\varphi)\right).10^6\right]^{1/2} \tag{1}$$

$$q(\varphi) = \frac{\rho_g}{6.n}.\int_{\alpha_0}^{\varphi} \frac{dq(\varphi)}{d\varphi} d\varphi \tag{2}$$

as from the inlet fuel pressure and the injector needle lift is calculated the approximate fuel pressure under the needle $p_g(\varphi)$ (phasing of the injection pressure and the reflected wave at the start of the needle lifting) and the current quantity of the fuel injected $q(\varphi)$ at a previously experimentally determined effective section of the injector nozzle $\mu F(\varphi)$. In Formula (1) the additional quantities are as follows: ρ_g- fuel density; φ – crankshaft rotation angle; $p_i(\varphi)$- current in-cylinder pressure; n - the crankshaft speed.

Prior to calculation, the input data was prepared according to the method described in [6], where the diagrams are represented in step of $0.1°\ CA$ (crankshaft rotation angle) and cubic spline transformation, ensuring continuity of the first and second derivatives, necessary for consecutive mathematical processing.

3. Experimental results and discussions

Engine operating modes simulating urban driving are explored. Transition processes cover engine speed ranges from 1500 to 3500 $min - 1$, which is a variation of engine speed relatively to the point of maximum torque of about 40 %. The load variation ranges from about 10 % to 100 % of the maximum torque value of the engine, and the individual tests are shown in table 1.

Table 1. Variation of engine load for individual tests relatively to the maximum torque of the engine.

No	Start condition	End condition
1	9.84 %	32.79 %
2	13.11 %	50.00 %
3	13.11 %	69.67 %
4	25.41 %	99.18 %

IOP Conf. Series: Materials Science and Engineering **614** (2019) 012008 doi:10.1088/1757-899X/614/1/012008

The duration of acceleration and time to complete the transition mode is about 6 s. This value corresponds to the average acceleration time of the vehicle in a given gear and partial loads in the presence of a speed limitation in urban environment.

Figure 1 shows the variation of the injected fuel per cycle q_c in a transient engine mode with an increase of the rotational speed rpm from 1500 to 3800 min^{-1} and a load increase of 12 to 40 N, where $Fdyn$ is the load of the engine dynamometer. The transient mode has a duration of about 6.2 s.

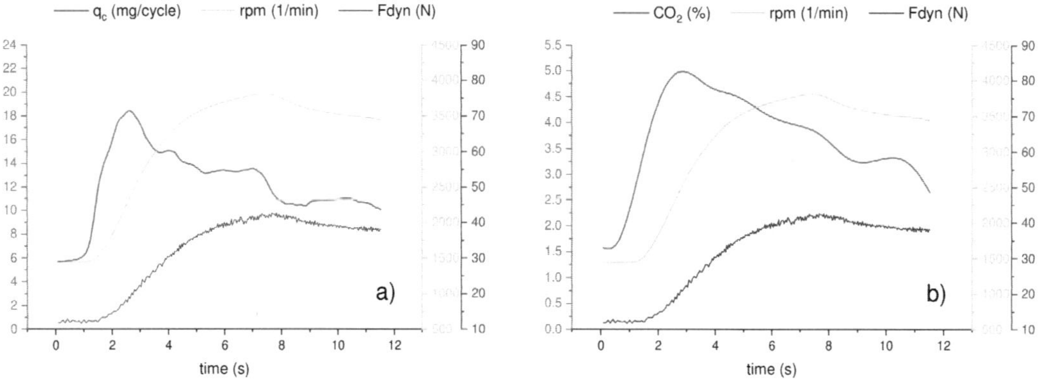

Figure 1. Injected fuel per cycle (a) and carbon emissions CO_2 (b) in transient engine mode with load up to 40 N.

At the beginning of this mode the engine runs at low load and fuel consumption is 5.8 $mg/cycle$. In the first phase of the transition regime, there is a sharp increase in fuel consumption and after about 2 s it reaches a local maximum of 18.38 $mg/cycle$ at a rotational speed of 2360 min^{-1}. Afterwards there is a decrease in consumption to reach the set speed of 13.3 $mg/cycle$. Once the load and speed regime has been established, the injected fuel per cycle is stabilized at 10.3 $mg/cycle$. Similar tendency is observed with the carbon emission results shown in figure 1. Minor differences in graphs are due to the damping effect of the total exhaust gases in the engine exhaust manifold when carbon dioxide data is recorded.

The amounts of hydrocarbons HC and nitrogen oxides NOx measured before the catalyst are shown in Figure 2.

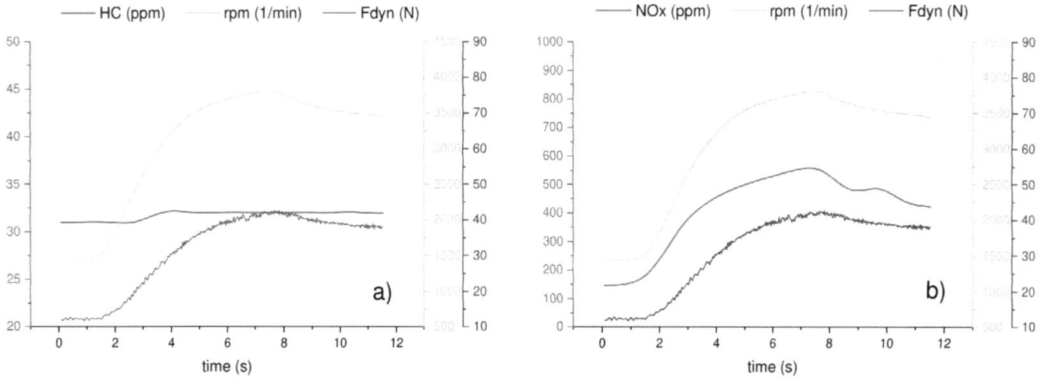

Figure 2. HC (a) and NOx (b) emission values for transient engine operation with load up to 40 N.

The HC values are not significantly affected by the parameters of the transition process. The behaviour of nitrogen oxides follows the increase in engine load and the fuelling behaviour of the engine. It can be seen from figure 2b) that in the acceleration mode, their values also increase beyond those of the established mode after reaching the set load. The maximum values are recorded at the moment when the maximum load and the rotation speed are reached, and thereafter they are reduced to those with established modes. The amount of nitrogen oxides emissions in the transition process compared to the stationary mode is about 130.9 %.

Figure 3 shows the results for transient operating modes of the engine with a 50 % increase in load to 60 *N* - figure 3a), and a 70 % increase in load to 85 *N*, figure 3b, relative to the initial load.

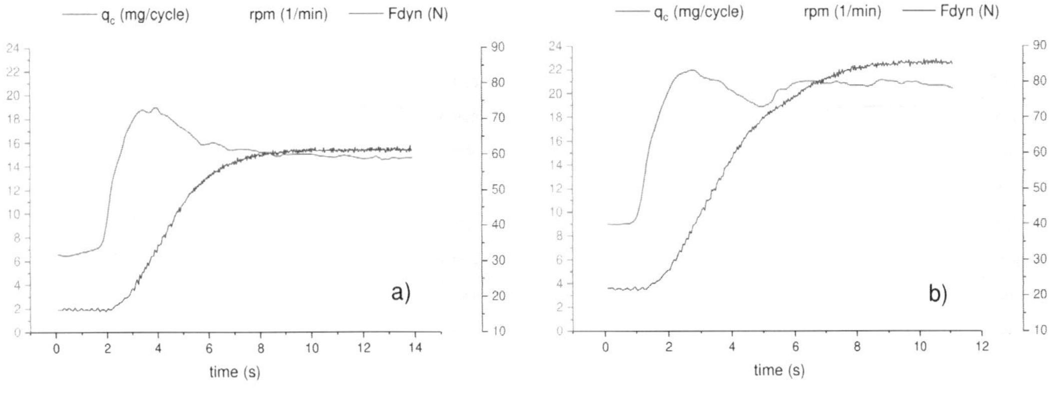

Figure 3. Injected fuel per cycle in transient engine operation with load up to 60 *N* (a); and up to 85 *N* (b).

The tendency to increase fuel delivery per cycle in the first phase of the transitional regime with a subsequent decrease in the established one is maintained. At higher final load values, the difference between the recorded maximum of the cyclic fuel portion in the transition and the established regime decreases. With a continuous increase in the final load mode, the local maximum disappears and the fuel delivery aims to reach that of the established mode.

Emissions of nitrogen oxides follow the increase in engine load. There is some lag, probably due to the lower thermodynamic conditions of the environment in which the working process develops in the first phase of the transition regime, figure 4.

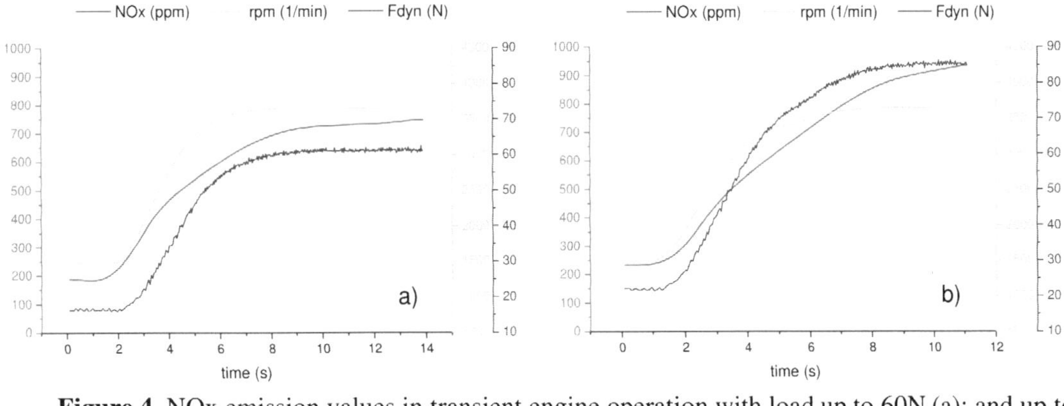

Figure 4. NOx emission values in transient engine operation with load up to 60N (a); and up to 85N (b).

Immediately after the start of the transition process, heat exchange with the walls of the combustion space is significant. This prevents the development of high temperatures in the combustion chamber and the generation of higher emissions of nitrogen oxides in the first phase. The regime with increasing load leads to a gradual increase of temperature of the combustion chamber and to higher maximum temperatures of the working cycle, with consequent impact on the emissions of nitrogen oxides. The results for NOx emissions at discussed loads of the engine dynamometer plus a full load curve are summarized in figure 5.

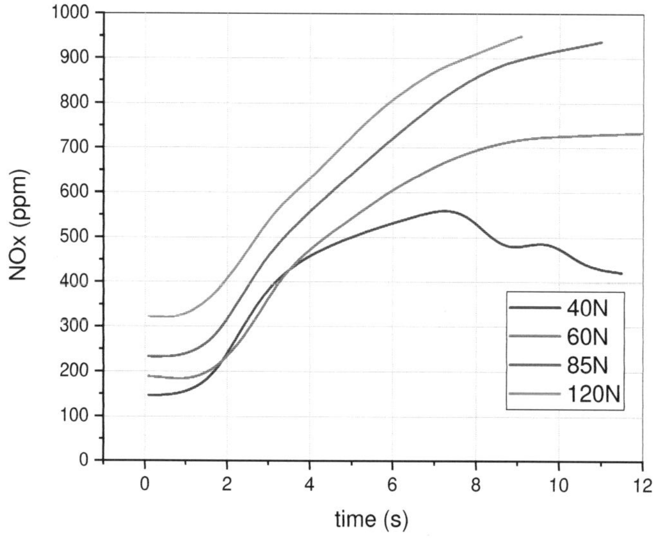

Figure 5. NOx emission values in transient engine operation with varying load.

4. Conclusion

Driving in urban areas is dominated by transient modes of operation. In such environment automotive engines operate with varying loads, frequent gear shifts and depending on the duration of the transition process, can be approximated with engine speed characteristics with partially open throttle.

The direct injection engine tests have shown the presence of phases of the transient modes in terms of fuel consumption and the content of harmful components in the exhaust gases. These phases are more pronounced in transient processes at engine operating modes in the low and medium load range. During the first phase at low speeds there is a rapid increase in the injected fuel per cycle as the engine tries to increase the kinetic energy of the vehicle and the rotating parts of the transmission. Together with increased fuel consumption, carbon dioxide and nitrogen oxide emissions are also increased. Next there is a drop in fuel delivery, which results from the reduced rate of rise of the rotation speed and gradually reaching the set load of the dynamometer. The third phase is related to the settlement of engine parameters to steady mode. By changing the mode in the area of high loads, the transient mode is characterized by the operation of the engine with wide-open-throttle speed characteristics.

In the first phase of the transient process, nitrogen oxides have a local maximum only at low loads. In such mode, after reaching the established parameters, as a result of reduced fuel delivery, the amount of nitrogen oxides also decrease. In all other modes, their amount is continuously increased as a result of improved thermodynamic combustion conditions in the engine cylinder.

By increasing the duration of the transition mode and creating conditions for the engine to work with part-open-throttle speed characteristics, there is a potential for reducing fuel consumption when driving in urban areas and reducing emissions of toxic components in exhaust gases.

References

[1] Disch C and Heiko K 2014 Untersuchung der Gemischbildung, Verbrennung und Schadstoffentstehung im transienten Betrieb an einem Ottomotor mit strahlgeführter Direkteinspritzung *Projektkennblatt der Deutschen Bundesstiftung Umwelt* (Karlsruhe: Institut für Kolbenmaschinen).

[2] Rakopoulos C and Giakoumis E 2009 *Diesel Engine Transient Operation* (London: Springer-

Verlag).
[3] Gerstle M 1999 Simulation des instationären Betriebsverhaltens hochaufgeladener Vier und Zweitakt Dieselmotoren (Hannover: Universität Hannover).
[4] Davis P and Peckham M 2007 The Analysis of Gasoline Transient Emissions Behaviour Using Fast Response Gas Analysers *SAE Paper 2007-26-015*.
[5] Tiedemann T 2018 Messung und Rechnung instationärer Betriebszustände am Beispiel eines abgasturboaufgeladenen Ottomotors (Hamburg: Helmut-Schmidt-Universität).
[6] Ivanov N and Ivanov Z 2002 Method for preprocessing of experimental data from internal combustion engines testing, registered with ADC, *Proc. of Scientific-technical union of mechanical engineering* **2** (Varna: NTS).

IOP Conf. Series: Materials Science and Engineering **614** (2019) 012009 doi:10.1088/1757-899X/614/1/012009

Study of the potential for reducing CO_2 emissions from road transport in Bulgaria

R Hristov[*], S Stefanov

Faculty of Mechanical Engineering and Technologies, Technical University of Varna, 1 Studentska Street, Varna 2010, Bulgaria

[*] corresponding author: rosen.hristov@tu-varna.bg

Abstract. Environmental issues take an extremely important place in the life of modern society and the measures which individual states and organisations take to reduce the harmful effect of transport are essential. The report examines the legal framework and the actions taken by one of the European Union states - Bulgaria. As regards CO_2 emissions, comparative results from a test of internal combustion engine fuelled with Petrol, LPG and CNG are shown. A survey was conducted to obtain the opinion of vehicle drivers on environmental issues and on introduction of cleaner vehicles - hybrid and electric vehicles. The results show that the reason for small electric vehicle sales is the lack of recharging points and the high price. The production, life and recycling of batteries are the other major issues noted in the survey. At present, the predominant production of electricity from thermal power stations using coals as a fuel does not give an advantage to electric transport, except to improve ecology in urban environment.

1. Introduction

In recent years, strategies and specific actions to reduce the emissions from vehicles have been taken in Europe, Asia and America. Publications that some countries will prohibit the registration of ICVs after a certain time period - in Norway, for example, this is the year 2025 - have appeared. In particular, there are studies on CO_2 reduction in certain regions and countries [1,2] related to the effect of demographic indicators on the use of electric vehicles. A reported difference of 30 - 40 % between official CO_2 values and real-world estimates was found [3]. There was made simulation models that assumptions were used about car ownership and volume, electric cars, fuel efficiency and electricity mix [4].

Against this background, Bulgaria, which has a fleet of over 15-20 year old vehicles, has the very difficult task to change in practice the pollution of environment by vehicles. The improvement of transport infrastructure does not always have a positive effect on the reduction of transport emissions; very often it provides opportunities for higher speeds and higher traffic. The replacement of urban transport with more environmentally friendly one is a good solution, but better organisation and incentives for its wider use are also needed.

JATO Dynamics Agency published a ranking (figure 1), which ranked European vehicle makes depending on the average CO_2 emissions that the vehicles in the portfolio released in 2018. With no surprises, Tesla is at the top with a score of 0 grams per kilometre, which along with Smart (89.8 grams of CO_2 per kilometre) occupies the only positions in the group 0-90 grams of CO_2. The next group - 90-110 grams of CO_2 per kilometre - includes four makes - Toyota (99.9), Peugeot (107.7), Citroen (107.9) and Renault (109.1), which form 22% of the total new vehicle market in Europe. The third and largest group - 110-130 grams of CO_2 per kilometre - includes 21 makes, which total market share is 67%. It is clearly apparent that when the limit of 95 grams per kilometre of CO_2 is introduced in 2021, almost no vehicle company will be able to achieve it. This is precisely the reason why electric vehicles and vehicles with hybrid drives are being developed to reduce the overall emissions of the company. At this stage, these types of vehicles still have a high price and their purchase needs to be supported by governments and municipalities.

1 Tesla	0.0	0 - 90 g/km 2017: 0.2%
2 Smart	89.8	
3 Toyota	99.9	90-110 g/km: 22% 2017: 22%
4 Peugeot	107.7	
5 Citroen	107.9	
6 Renault	109.1	
7 Nissan	110.6	110-130 g/km: 67% 2017: 73%
8 DS	114.2	
9 Suzuki	114.2	
10 Skoda	116.7	
11 Lancia	116.7	
11 Seat	116.9	
15 Volkswagen	118.8	
16 Fiat	119.2	
16 Mitsubishi	119.5	
16 Mini	119.9	
17 Kia	120.4	
18 Dacia	120.8	
19 Lexus	122.5	
20 Hyundai	125.5	
21 Ford	125.7	
22 Opel/Vauxhall	125.6	
23 Honda	127.0	
24 Audi	127.6	
25 BMW	128.9	
25 Alfa Romeo	128.9	
27 Volvo	150.0	
28 MG	151.7	130-160 g/km: 9% 2017: 4%
29 Mazda	155.2	
30 Mercedes	159.6	
31 Abarth	141.7	
32 Jeep	142.5	
33 Jaguar	147.8	
34 Infiniti	142.9	
35 Subaru	160.5	160+ g/km: 1.9% 2017: 0.6%
36 Ssangyong	164.8	
37 Land Rover	166.7	
38 Porsche	185.0	
39 Lada	186.8	
40 lotus	208.9	
41 Alpina	209.0	
42 Maserati	225.4	
43 McLaren	249.6	
44 Cadillac	255.8	
45 Chevrolet	255.8	
46 Aston Martin	261.8	
47 Bentley	275.2	
48 Ferrari	283.1	
49 Lamborghini	325.2	
50 Rolls-Royce	527.6	

Figure 1. JATO Dynamics Agency ranking European vehicle makes depending on the average CO_2.

2. Regulatory Basis and Future Actions

Key documents for the Transport sector in Bulgaria are:

- Integrated Transport Strategy for the period until 2030
- National Policy Framework for Development of Alternative Fuels Market in the Transport Sector and Deployment of the Relevant Infrastructure.

Some of the strategic priorities in transport development are:

- Effective maintenance, modernisation and development of transport infrastructure
- Reducing fuel consumption and increasing energy efficiency in transport
- Ensuring high-quality and accessible transport in all regions of the country
- Limiting the negative effect of transport on the environment and human health
- Increasing the security and safety of transport system.

The overall objective of the framework is to create a favourable environment for wider application of alternative fuels and drive mechanisms in transport sector and to determine conditions which are comparable to those of other developed countries of the European Union. In the long-term (the period after 2030), the objective is to develop electromobility, to use wider natural gas as a standard fuel, and to move forward hydrogen technology from the research / development phase.

In May 2016, a pilot scheme was launched for the first time in the country. It provided partial funding of the funds needed to purchase electric vehicles or hybrid electric vehicles for public institutions, thus optimising fuel costs and saving greenhouse gas emissions. The funding amounts to BGN 1 million.

Over 76% of the recharging points in European Union are located in 4 countries - Germany, the Netherlands, England and France (figure 2). Bulgaria takes one of the last places in Europe regarding the number of electric vehicle recharging points [5].

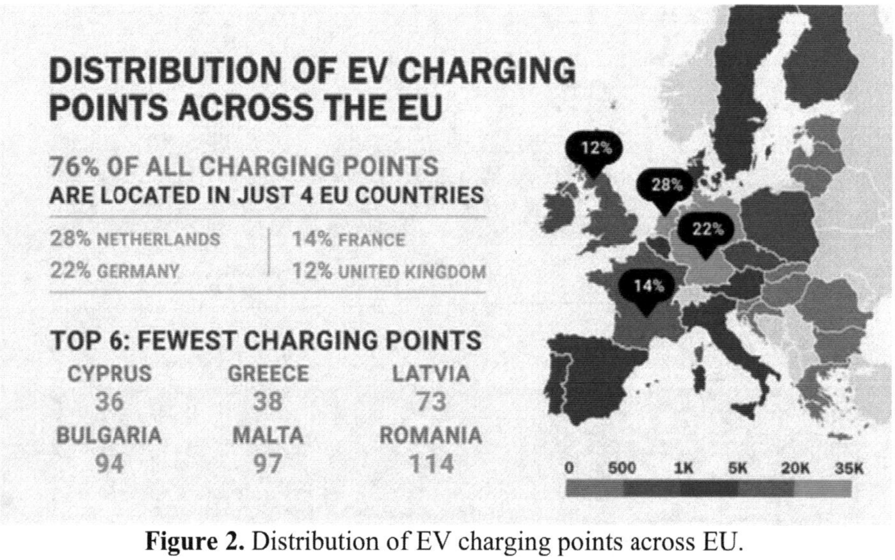

Figure 2. Distribution of EV charging points across EU.

All charging stations are located within the urban areas. No charging station has been installed outside urban areas yet. At municipality level, the option to install charging stations for electric vehicles with normal and high power at key points, which allow for normal service of electric and hybrid drive vehicles, could be explored. The study should cover the availability of spare capacity or the ability to provide such capacity in the appropriate locations without major infrastructure investments. It should be conducted jointly with the electricity distribution company operating on the territory of the respective municipality. An important parameter is also the energy mix: that is how electricity is produced [6].

The use of gases provides the old petrol vehicle fleet with an opportunity to significantly reduce the harmful exhaust emissions. The lower operating costs stimulate the drivers to shift their focus to this type of fuel. This is more relevant for liquefied petroleum gases and applies limited to compressed natural gas (CNG). The reason is the more developed LPG charging station system and the lower price for converting the vehicles. Despite the lower excise duty on gases, the lack of higher competition results in higher price levels of compressed natural gas. Figure 3 shows the results of our study of an

IOP Conf. Series: Materials Science and Engineering **614** (2019) 012009 doi:10.1088/1757-899X/614/1/012009

older generation petrol engine fuelled with petrol, LPG and CNG. It is clearly apparent that CO_2 emissions can be significantly reduced without too many efforts in the operation with natural gas. Biogases are not very common in Bulgaria as a fuel for internal combustion engines, since they have to undergo intermediate treatment to remove impurities. There are developments of biogas fuel cells that could be used in electric vehicles.

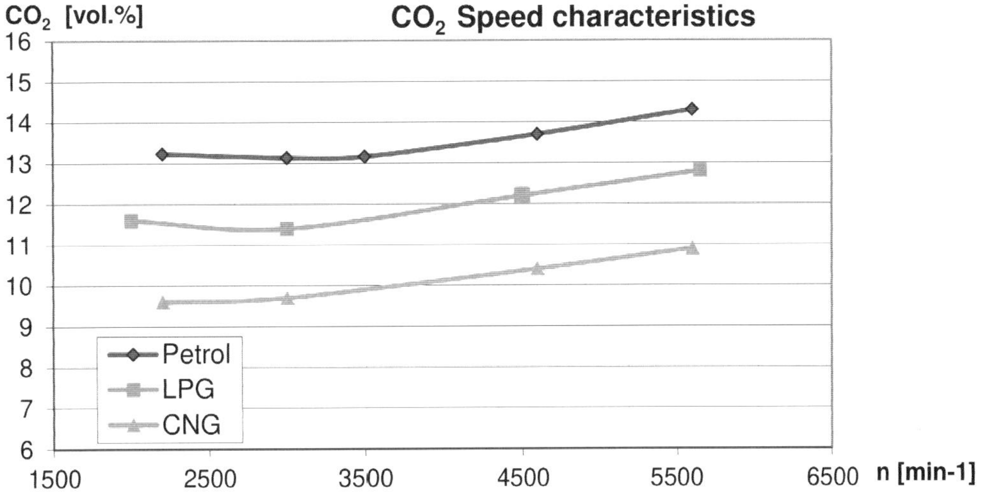

Figure 3. Comparison of emissions from ICE with different types of fuels.

3. Survey results

A survey was conducted to obtain the opinion of vehicle drivers on the environmental issues and pollution from vehicles using internal combustion engines. It gives us information about the reason for the small distribution of electric and hybrid vehicles. Around 76% of the total 500 respondents are men, 24% are women, and 84% are young people and middle-aged people up to the age of 45 (figure 4). Sixty percent of them have diesel vehicles, two percent have hybrid vehicles and no one is an owner of electric vehicle (figure 5). This is not surprising, since in Bulgaria there are less than 800 electric vehicles. The majority of the vehicles are over 15 years of age (56%). When asked about the desired new vehicle, most of the respondents answered that they wished to have a petrol vehicle (60%). Electric vehicle took the last place (4% of the respondents) (figure 6).

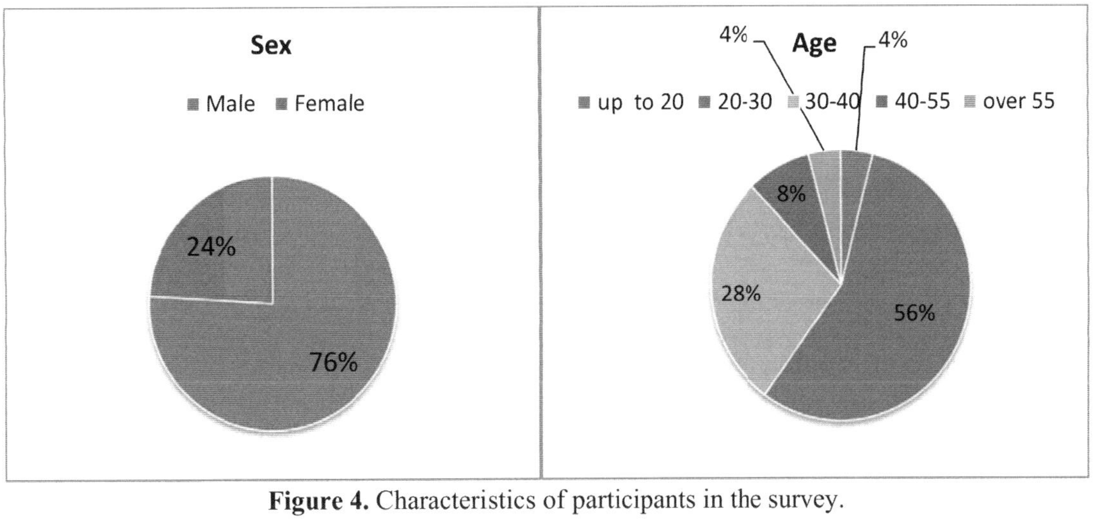

Figure 4. Characteristics of participants in the survey.

IOP Conf. Series: Materials Science and Engineering **614** (2019) 012009 doi:10.1088/1757-899X/614/1/012009

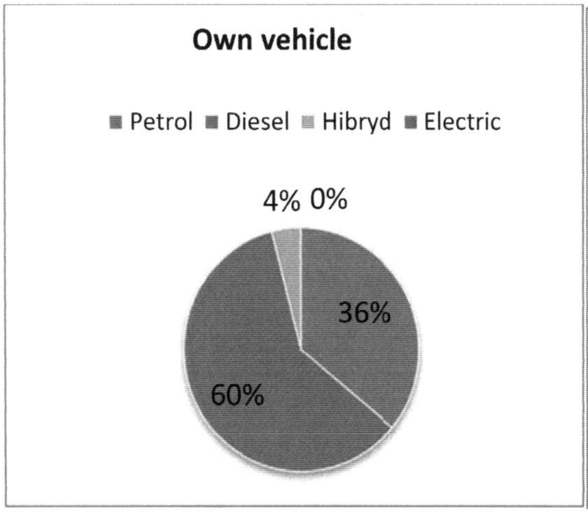

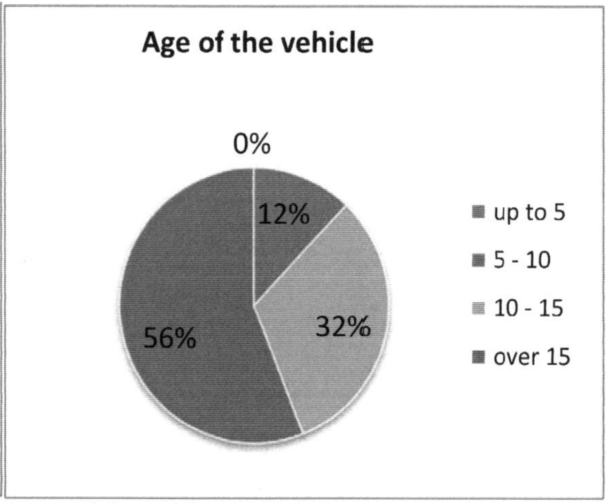

Figure 5. Type and age of the vehicles.

Walking is a preferred way of movement in urban environment, followed by urban transport and bicycle (figure 7). As a pollutant, road transport is ranked first by 8% of the respondents, second - by16% of the respondents, third- by 36% of the respondents, and fifth - by the 40% of the respondents (figure 8). In Bulgaria, the import of second-hand vehicles is particularly important, and in particular the concern that all old diesel vehicles produced in Germany, Austria, Italy, etc. will be directed to our country.

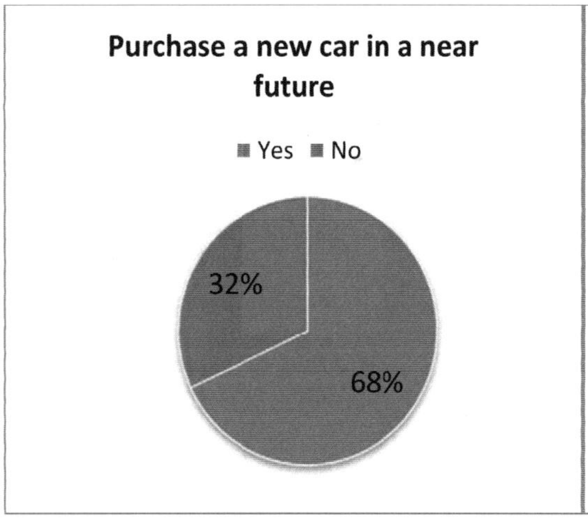

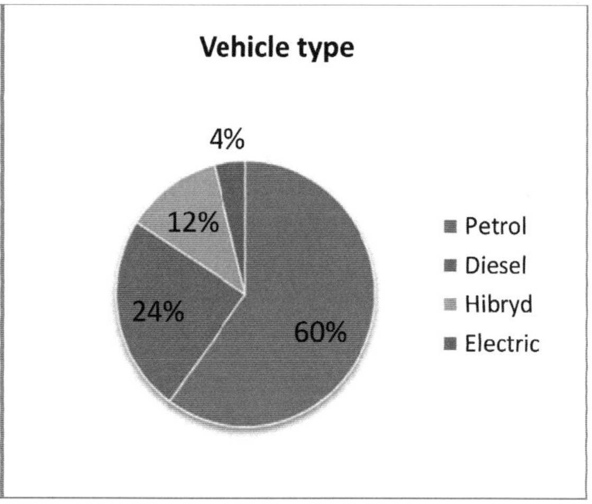

Figure 6. The probability of buying a new car and a desired type.

When asked whether the import of vehicles after introduction of Euro 5 would deteriorate the environmental situation, 52% of the respondents answered "No" (figure 8).

The answers to the question „Which is the reason for the small number of electric vehicles in Bulgaria?" are ranked in the following way:

1. Underdeveloped charging station network.
2. High price.
3. Doubt in the reliability and resource.
4. Lack of government subsidies.
5. Insufficient commitment of people to environmental issues.
6. Difficult recharging at home (lack of garage or parking space).

IOP Conf. Series: Materials Science and Engineering **614** (2019) 012009 doi:10.1088/1757-899X/614/1/012009

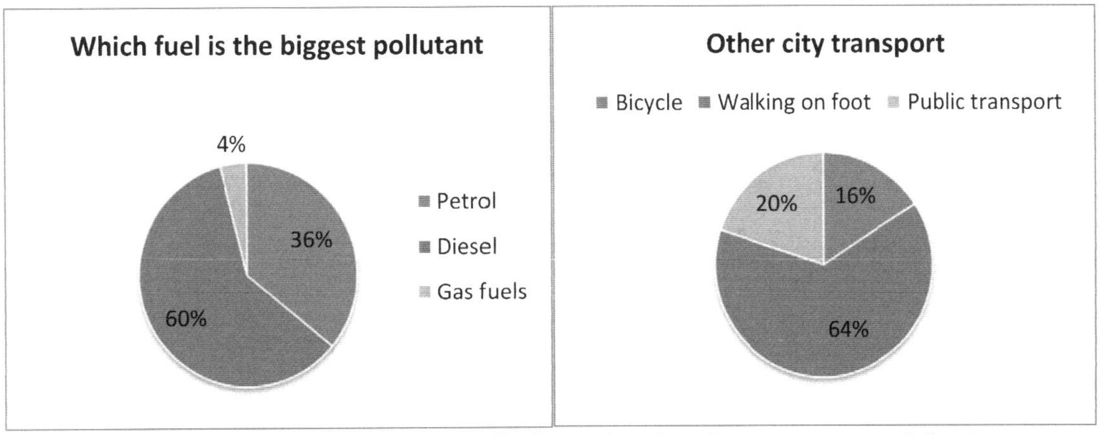

Figure 7. Which fuel is the most polluting and preferred way to get around the city.

The answers to the question „Which, in your opinion, is a disadvantage to use electric vehicles at this stage?" are ranked in the following way:

1. Energy-intensive and environmentally unfriendly battery production.
2. Problems with recycling of end-of-life batteries.
3. Production of electricity mainly by thermal power plants .
4. Future increase in electricity prices.

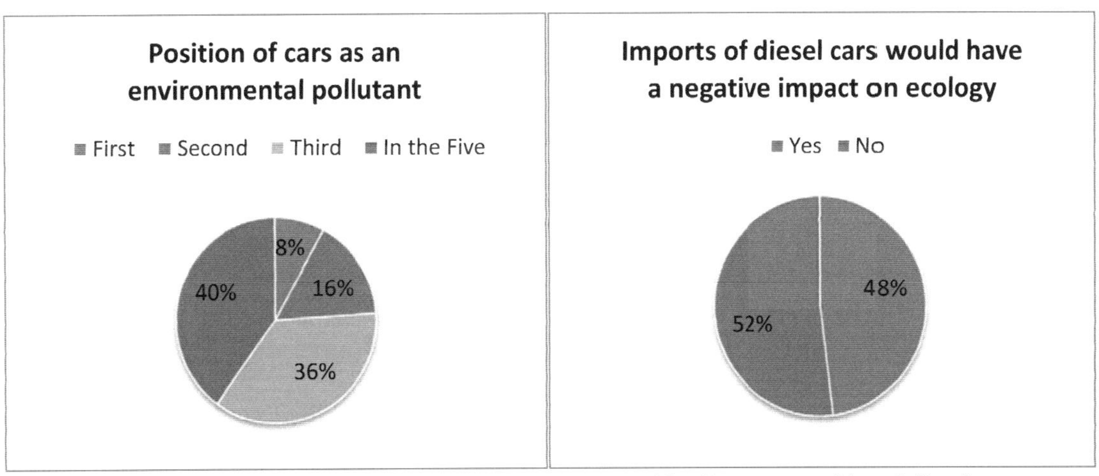

Figure 8. Opinion on vehicle pollution and specifically on second-hand diesel.

Conclusion

To actually reduce CO_2 emissions from transport in Bulgaria, the state needs to take some more serious measures to improve the age of vehicles fleet, to subsidise the purchase of electric vehicles and to make efficient use of the EU funds. The continuation of subway network in the city of Sofia and the replacement of diesel buses with electric buses and electric buses are good steps forward. Municipalities can support, using different incentives, the development of charging stations for electric vehicles, in order to eliminate one of the main problems in the opinion of the people - poor distribution and insufficient number of charging stations. This is very important, since most of the vehicle owners do not have a garage or permanent parking space and therefore they cannot rely on home recharging. In general, the use of gases is widespread, and only in this direction reduction of CNG prices may be sought.

Acknowledgments
This work received funding from the Technical University of Varna by project KD1 with a project leader Assoc. prof. PhD eng. R. Hristov.

References
[1] Sergey P, Chen Y, Karplus V, Kishimoto P, Reilly J, Löschel A, Graevenitz K and Koesler S 2018 Reducing CO2 from cars in the European Union *Transportation* **45** pp 573-95.
[2] Sovacool B, Kester J , Noelb L, Zarazua G 2018 The demographics of decarbonizing transport: The influence of gender, education, occupation, age, and household size on electric mobilitypreferences in the Nordic region *Global Environmental Change* **52** pp 86 – 100.
[3] Fontaras G, Zacharof N, Ciuffo B 2017 Fuel consumption and CO2 emissions from passenger cars in Europe Laboratory versus real-world emissions *Progress in Energy and Combustion Science* 60 pp 97 – 131.
[4] Vleugel J and Bal F 2018 The impact of a CO2 reduction target on the private car fleet in the Netherlands *WIT Transactions on Ecology and the Environment* **215** pp 109 – 112.
[5] ACEA 2018 Making the transition to zero-emission mobility www.acea.be/publications/article/study-making-the-transition-to-zero-emission-mobility.
[6] Noga M and Juda Z 2018 Energy efficiency and equivalent CO2 emissions of a light-duty electric vehicle depending on driving distance *IOP Conf. Ser.: Mater. Sci. Eng.* **421** 022023.

IOP Conf. Series: Materials Science and Engineering **614** (2019) 012010 doi:10.1088/1757-899X/614/1/012010

General Motors 6T automatic gearboxes - problems and methods for their removal

Stefanov S*, Hristov R

Faculty of Mechanical Engineering and Technologies, Technical University of Varna, 1 Studentska Street, Varna 2010, Bulgaria

* corresponding author: stefanov_48@abv.bg

Nowadays, with increased environmental requirements to improve fuel economy, it is important to work with both the ICE and the other elements of propulsion. This report describes how to operate automatic gearboxes and specially General Motors 6T. They are very widespread throughout the world and are used by many automotive manufacturers. The deficiencies encountered during the operation and the reasons for this are discussed. There was considered the opportunities for diagnosis, emerging operational problems and the options for their removal. It is suggested to replace parts exchange details of the construction in order to improve performance and reliability. The solenoids that control the switching are ON / OFF type or pressure regulators (with a progressive or regressive characteristic). A stand test of solenoids - pressure regulators has been done under cold and warm conditions. The test results are shown in tabular and graphical form. Using this method reduces the repair time of an automatic gearbox.

1. Introduction

Automatic transmissions in vehicles have been known since the middle of the last century. During that time they were hydraulically controlled. In the beginning, the gears were two or three, and then their number increased. In modern automatic gearboxes the number of gears is between six and ten. The increased number of gears improves the economy of vehicle and travel comfort. In addition, the engine runs in its optimal modes, which leads to release of lower emissions in the atmosphere. To achieve optimal alignment, the engine and gearbox operation helps the electronic control that presents in every modern automatic transmission.

This report will focus on the General Motors 6T automatic gearboxes. We will consider the opportunities for diagnosis, emerging operational problems and the options for their removal. There is also an opportunity for analytical research, but the best is the experimental determination of the parameters [1, 2].

2. Explanation

The objectives of this report are:

1. To clarify the principle of operation of these gearboxes.
2. To identify the failures and the reasons for their occurrence.
3. To determine the condition of electronic pressure regulators in the control.

2.1. Principle of gearbox operation

The 6T series automatic gearboxes have been developed in several versions to cover the various power ranges of the engines which they will work with. These are 6T30E, 6T35E, 6T40E, 6T45E, 6T50E, and 6T70 / 75E. Their installation started in 2007. There are two transmission options: front-wheel drive and all-wheel drive (4x4). Two versions of the models have been developed: generation 1 and generation 2. From 2014, generation 3 version also comes into operation.

This transmission is quite widespread. It is mounted to Opel Astra, Chevrolet Captiva, Chevrolet Cruz, Chevrolet Malibu, Buick laCrosse, Ford, etc.

The manufacturer has indicated the following advantages of this gearbox design:

1. Minimum overall dimensions.
2. Improved fuel economy.
3. Improved gear shifting.
4. New design of the converter (hyperelliptic converter), which saves the overall dimensions.
5. Unique input shaft.
6. New oil pump design with reduced noise level and improved efficiency.

These transmissions are mainly produces in Korea, China and Mexico. Its improvement continued steadily [3].

The diagram of this transmission is shown in Figure 1. It consists of five multi-disc clutches and one-way clutch (OWC), which are used to operate six forward drive gears and a reverse gear.

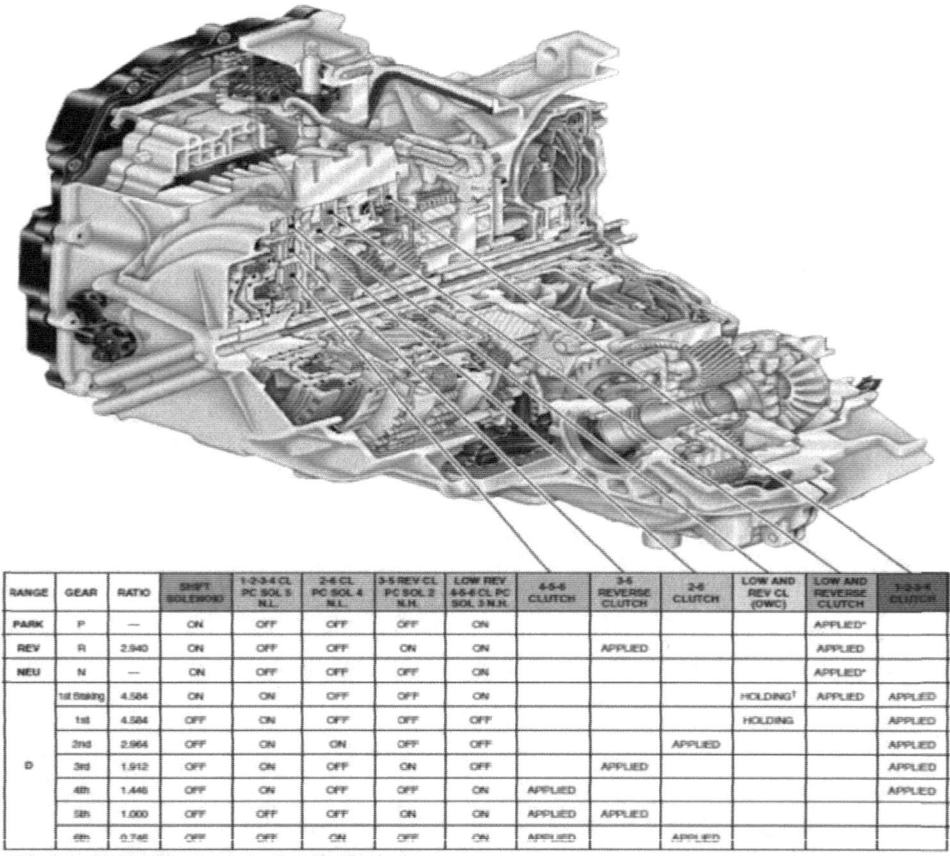

RANGE	GEAR	RATIO	SHIFT SOLENOID	1-2-3-4 CL PC SOL 5 N.L.	2-6 CL PC SOL 4 N.L.	3-5 REV CL PC SOL 2 N.H.	LOW REV 4-5-6 CL PC SOL 3 N.H.	4-5-6 CLUTCH	3-5 REVERSE CLUTCH	2-6 CLUTCH	LOW AND REV CL (OWC)	LOW AND REVERSE CLUTCH	1-2-3-4 CLUTCH
PARK	P	—	ON	OFF	OFF	OFF	ON					APPLIED*	
REV	R	2.940	ON	OFF	OFF	ON	ON		APPLIED			APPLIED	
NEU	N	—	ON	OFF	OFF	OFF	ON					APPLIED*	
D	1st Braking	4.584	ON	ON	OFF	OFF	ON				HOLDING†	APPLIED	APPLIED
	1st	4.584	OFF	ON	OFF	OFF	OFF				HOLDING		APPLIED
	2nd	2.964	OFF	ON	ON	OFF	OFF			APPLIED			APPLIED
	3rd	1.912	OFF	ON	OFF	ON	OFF		APPLIED				APPLIED
	4th	1.446	OFF	ON	OFF	OFF	ON	APPLIED					APPLIED
	5th	1.000	OFF	OFF	OFF	ON	ON	APPLIED	APPLIED				
	6th	0.746	OFF	OFF	ON	OFF	ON	APPLIED		APPLIED			

NOTE: EFFECTIVE FINAL DRIVE RATIO – 6T40/45 – 3.17 & 3.87
NOTE: FOR SHIFT SOLENOID, "ON" – SOLENOID ENERGIZED (PRESSURIZED), "OFF" – SOLENOID DE-ENERGIZED (NO PRESSURE).
NOTE: FOR PRESSURE CONTROL (PC) SOLENOIDS, "ON" – PRESSURIZED, "OFF" – NO PRESSURE.
* APPLIED WITH NO LOAD.
† HOLDING BUT INEFFECTIVE.

Figure1. Diagram of 6T transmission.

Electrohydraulic control system called Mechatronics, which is located behind a plastic cover in the gearbox, controls the elements during gear shifting. The transmission control module (TCM) is also mounted there, as is the case with almost all modern automatic gearboxes.

Figure 2 shows the mechatronics electronic part and solenoids.

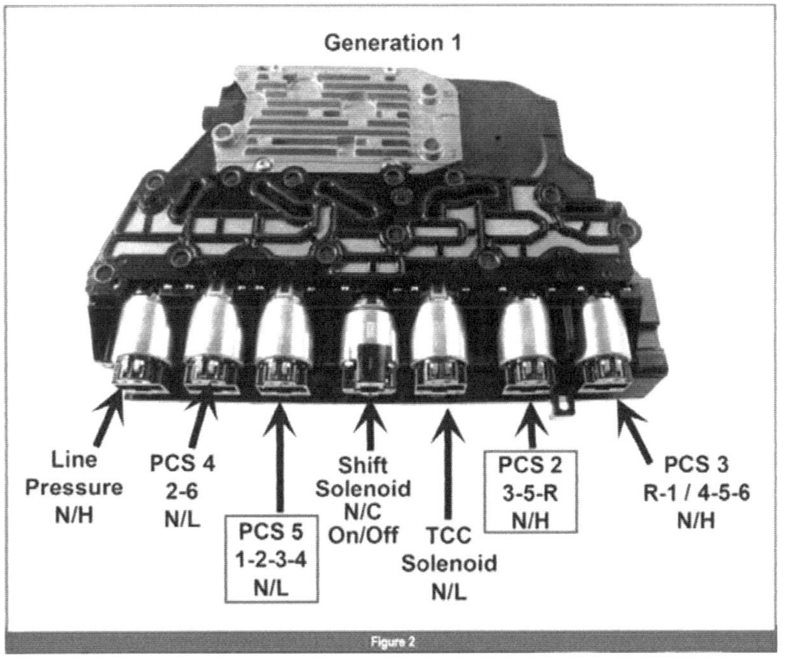

Figure 2. Solenoids and TCM.

2.2. Faults in the operation of this type of transmission and reasons for their occurrence.

The following defects are observed during the operation of 6T gearboxes:

- broken primary shaft;
- worn bearing bushes which house the primary shaft and worn other bearing bushes (10 in total);
- damaged drum of clutches (4-5-6) and (3-5-R);
- restraining OWC;
- damaged splines of stator oil pump, which houses the one-way clutch of converter reactor;
- worn clutches;
- pressure leakage through the pistons that press the clutches;
- worn Lock up in the converter;
- problems in mechatronics.

Since gearboxes of this type are equipped with lots of electronic elements, it is essential to test the vehicle with a tester to verify whether any failures are recorded.

If electrical problems are found, they should be removed and appropriate follow-up actions should be subsequently taken. For more accurate diagnostics, sensors can be placed in the system so that malfunctions can be detected and isolated accurately [4, 5].

After dismantling the gearbox, the components must be carefully checked for defects and the reason for their occurrence should be sought.

Common defects in these gearboxes are: worn bearing bushes and problem in the drum of the two clutches (4-5-6) and (3-5-R) – figure 3.

The bearing bushings are examined, measured and, if necessary, replaced.

Since the two clutches (4-5-6) and (3-5-R) are supplied with pressure through the drum shell, the pressure leakage through the seal adversely affects their operation.

The pistons that engage the clutches also cause problems - fig. 4. They are made of embossed plate on which rubber seal is vulcanised so as to obtain a lip seal on the opening and shaft.

Figure 3. Drum of clutches (4-5-6) and (3-5-R).

Figure 4. Piston.

The pistons that engage the clutches also cause problems - fig. 4. They are made of embossed plate on which rubber seal is vulcanized so as to obtain a lip seal on the opening and shaft. The curing of these sealing surfaces leads to hydraulic leaks and loss of pressure.

When dismantling and repairing these gearboxes, it is recommended to replace some of the parts with parts with updated design. Such improved variants are offered for the drum (4-5-6) and (3-5-R): corrugated end plate, clutch slip washer (3-5-R), etc. depending on the modification.

2.3. Faults in the Control

Gearbox control consists of hydraulic and electronic part.

The hydraulic unit must be cleaned and checked for wear-and-tear. In addition, the main hydraulic regulator and TCC control valve located in the oil pump housing must be checked.

The solenoids that control switching are seven in number. One of them (shift solenoid) is ON / OFF type and is easy to check. The other six solenoids are pressure regulators. They are two types: three of them have progressive characteristic and three of them - regressive characteristic. These solenoids are tested on a stand. It is recommended to check their characteristics under cold and warm conditions.

Stand tests of solenoids were conducted on a stand specially designed and manufactured by the author [6]. 8 solenoids - 4 solenoids with progressive characteristic and 4 solenoids with regressive characteristic - were used in the experiment. The characteristics were determined at supply pressure of 10 [bar] and frequency of 300 [Hz]. The operating frequency and duty cycle are set by electronic system specifically designed for that purpose. In parallel to the signal produced for the solenoid, signal control oscilloscope is connected. The regulated pressure given in the two tables is measured in [bar].

2.3.1. Stand experiments. The test results are shown below in tabular and graphical form. Table 1 and Figure 5 show the pressure in the pressure regulators at normal temperature and Table 2 and Figure 6 at operating temperature.

IOP Conf. Series: Materials Science and Engineering **614** (2019) 012010 doi:10.1088/1757-899X/614/1/012010

Table 1. Cold oil.

Duty[%]	Reg. 1	Reg. 2	Reg. 3	Reg. 4	Reg. 5	Reg. 6	Reg. 7	Reg. 8
0	7.88	7.97	6.48	7.24	0.71	0.40	0.04	0.07
4	7.50	7.48	6.30	6.99	0.95	0.71	0.04	0.07
9	6.46	6.44	4.95	5.57	1.80	1.55	1.75	0.13
15	5.11	4.91	3.57	3.95	3.04	3.00	3.06	0.24
20	3.88	3.66	2.57	3.17	3.86	4.13	4.04	0.38
25	2.80	2.55	1.58	2.35	4.99	5.02	5.26	0.47
29	1.78	1.44	0.98	1.73	6.17	6.08	6.13	0.55
34	0.91	0.60	0.55	1.29	7.04	7.01	6.97	0.62
40	0.42	0.04	0.27	1.02	7.84	8.12	7.72	0.71
45	0.07	0.04	0.22	0.82	8.46	8.55	8.08	0.73
50	0.04	0.04	0.18	0.69	8.99	9.32	8.50	0.80

Table 2. Operating temperature.

Duty[%]	Reg. 1	Reg. 2	Reg. 3	Reg. 4	Reg. 5	Reg. 6	Reg. 7	Reg. 8
0	7.48	7.46	3.11	5.28	0.80	0.80	0.75	1.35
4	7.24	7.15	3.11	5.24	0.89	0.95	1.11	1.35
9	6.33	6.57	3.06	4.51	1.13	1.40	1.69	1.35
15	5.26	5.33	2.62	3.53	1.93	2.26	1.84	1.35
20	4.04	4.28	2.42	2.75	2.71	3.17	3.22	1.35
25	2.89	3.22	1.78	2.09	3.73	4.20	4.59	1.35
29	2.49	2.29	1.24	1.42	4.75	5.19	5.73	1.35
34	1.66	1.42	0.84	1.02	5.97	6.33	6.66	1.35
40	1.00	0.64	0.55	0.78	6.97	7.44	7.64	1.35
45	0.49	0.24	0.38	0.67	7.90	8.35	8.50	1.35
50	0.07	0.04	0.24	0.55	8.77	9.35	9.46	1.38

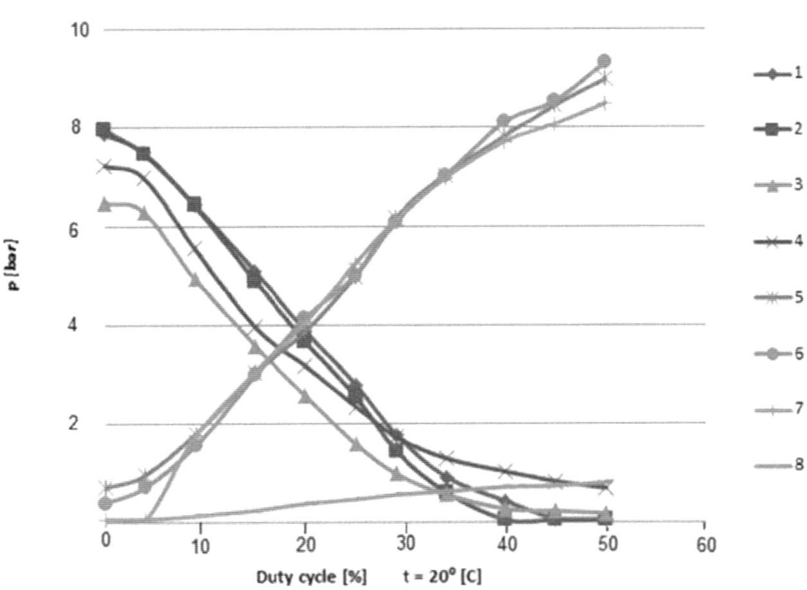

Figure 5. Characteristics - room temperature.

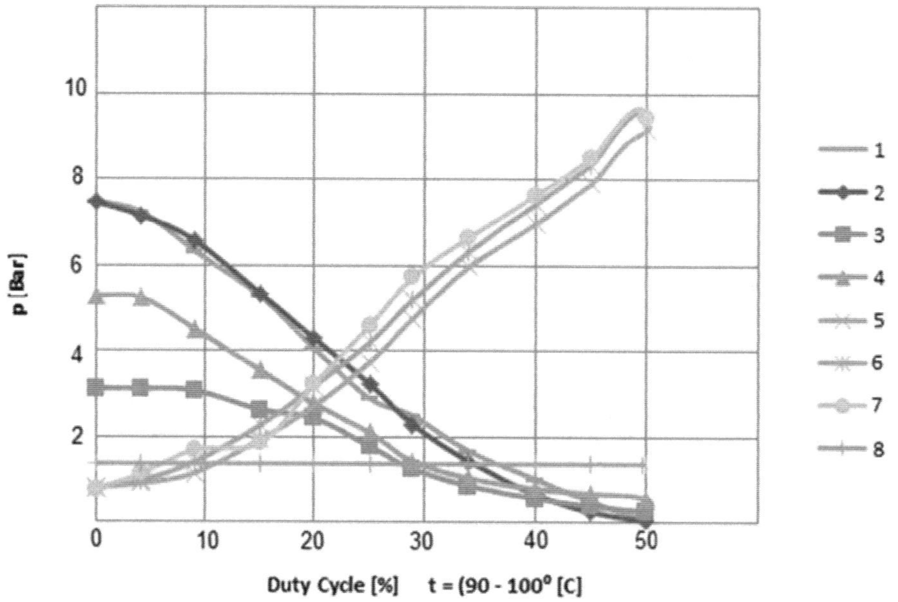

Figure 6. Characteristics - operating temperature.

2.3.2. Analysis of the Test Results. The experimental data and graphics show that under both cold oil and operating temperature conditions regulators 1,2,5,6 and 7 are in order. Regulators 3, 4 and 8 do not behave properly, with this being more pronounced at operating temperature.

3. Conclusion

The design of General Motors 6T automatic gearboxes is modern, compact and in line with the current trends for development of automatic gearboxes.

Some failures described in the report occur in the process of operation. They are successfully removed by replacing the damaged parts with improved ones, which are offered in the stores.

Electronic pressure regulators must be tested on a stand when problems during their operation occur and when the gearbox is repaired.

References

[1] Hongchao J, Wei W, Hongcai L and Qingdong Y 2018 Optimization of a pressure control valve for high power automatic transmission considering stability *Mechanical Systems and Signal Processing* **101** pp 182-96.

[2] Chen Q, Ahmed Q, Rizzoni G, Frisk E and Zhai H, Model-Based Fault Diagnosis of an Automated Manual Transmission Shifting Actuator 2015 *IFAC-PapersOnLine* **48-21** pp 1479-84.

[3] Clark K et al. 2015 General Motors Front Wheel Drive Seven Speed Dry Dual Clutch Automatic Transmission *SAE Int. J. Engines* **8(3)** pp 1379-90.

[4] Ahmed Q, Arasu M, Zhang J and Rizzoni G 2016 Sensors Installation Guide to Monitor Automatic Transmission Performance *IFAC-PapersOnline* **49-11** pp 736-41.

[5] Parham M, Jeevan M, Pratik P, Debangsu B and Raghunathan R 2015 Optimal Sensor Placement for Fault Diagnosis Using Magnitude Ratio *Industrial & Engineering Chemistry*

IOP Conf. Series: Materials Science and Engineering **614** (2019) 012010 doi:10.1088/1757-899X/614/1/012010

Research, **54 (38)** pp. 9369-81.
[6] Stefanov S 2012 *Proc. Int. Conf. EKO Varna (Varna)* vol. 19 (TU Varna) p147.

IOP Conf. Series: Materials Science and Engineering **614** (2019) 012011 doi:10.1088/1757-899X/614/1/012011

Modeling and investigation of a diesel engine with ethanol additives

S Iliev*, E Mitev

University of Ruse, Department "Engines & vehicles", 8 Studentska Str., Ruse, Bulgaria

*E-mail: spi@uni-ruse.bg

Abstract. Ethanol is a promising alternative fuel, due to its renewable origin based on biomass. Also, it has a lower carbon content than the diesel fuel, which contains oxygen. For this reason, ethanol provides a remarkable potential for particulate reduction in diesel engines. One-dimensional simulation of a single-cylinder diesel engine with the help of the GT-Power software has been done. The simulation is designed to test the characteristics of the ethanol-filled diesel engine at full load. The obtained results were compared with pure diesel fuel.

1. Introduction

The increased air pollution from vehicles has reached high level, especially in large cities where is observed an increase in the population and the number of vehicles respectively. Exhaust emissions usually come from the combustion of fossil fuels in vehicle engines. The use of additives from oxygen-containing fuels (methanol, ethanol, butanol, biodiesel, etc.) improves the environmental performance of vehicles, but their increased share leads to a deterioration in engine power performance [1]. For this reason, they are added in small percentages so far. Some of the major additives to improve fuel efficiency are organic compounds containing oxygen (oxygenates). Some oxygen-containing compounds are used as fuel additives such as methanol, ethanol, tertiary butyl alcohol and methyl tertiary butyl ether. At present times, ethanol is a promising fuel used in vehicles as an alternative to petroleum fuels [2, 3]. One reason for using ethanol is that it can be produced from natural products or waste materials, unlike petrol, which is a non-renewable fuel source.

2. Exposition

Various biofuel and alcohol add-on studies have been conducted globally to see their impact on exhaust emissions, fuel consumption and engine performance. In one of the studies, the author studied the physic-chemical properties of mixtures of ethanol and diesel fuel. He concluded that the addition of ethyl alcohol in diesel to 15% prevents segregation and increases the temperature of self-ignition [4]. In another report [5], the author concluded that the addition of ethanol reduces CO, SO_2 and soot emissions, although it has led to an increase in NOx emissions and a reduction in power of approximately 12.5% (with the addition of 10 % ethanol) and 20% (by addition of 15% ethanol). Another author has investigated the combustion characteristics of engines that use ethanol blended diesel blends with and without a cetane number enhancer. Its study shows that with the same average effective pressure and engine speed, ignition delay period, the heat released and the combustion duration are improved [7]. In another study conducted with a 5.9 liter diesel engine, torque was tested with two blends containing 10% and 15% ethanol. Engine power is reduced by about 8% with both fuel blends.

Figure 1 shows a single-cylinder diesel engine, developed using the GT-Power software.

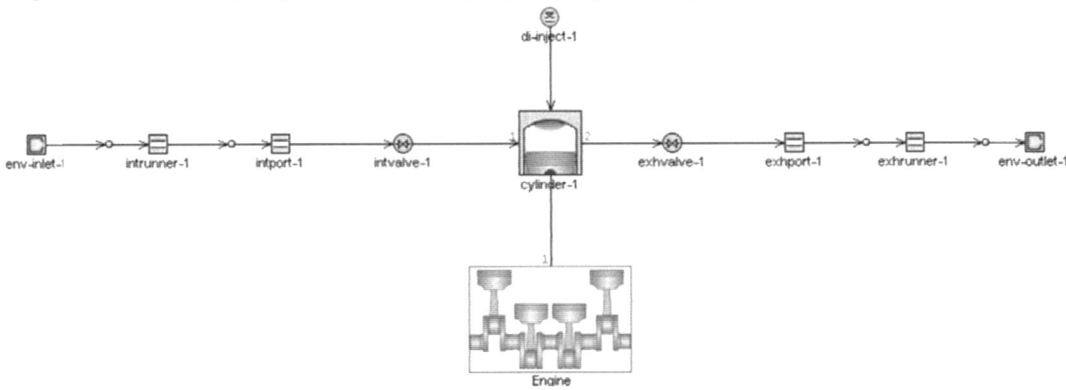

Figure 1. Single-cylinder diesel engine model

Table 1. Characteristics of the diesel engine

Description	Value
Fuel	Diesel
Engine type	Single cylinder
Piston diameter [mm]	87,5
Piston stroke [mm]	110
Volume [cc]	661
Compression ratio	17,5:1
Power [kW]	3,5
Maximum rotation speed [min^{-1}]	1500
Opening suction valve before TDC [^{0}]	4,5
Closing suction valve after BDC [^{0}]	35,5
Opening exhaust valve before TDC [^{0}]	35,5
Closing exhaust valve after BDC [^{0}]	4,5

The characteristics of the diesel engine used in the study are shown in tab. 2. Figure 2 shows the variation of the specific fuel consumption at different ethanol concentrations of 0% Ethanol (E0), 5% Ethanol (E5), 10% Ethanol (E10), 15% Ethanol (E15). The study was performed at a engine speed of 1500 min-1. The results obtained show that with an increase in the ethanol concentration in the mixture (E5 - E15) the specific fuel consumption increases, with a maximum specific consumption being obtained in mixtures with 15% ethanol E15. The lower ethanol calorific value of 26.8 compared to pure diesel fuel 43 (Table 1) increases the fuel consumption to obtain the same amount of energy. Also, the low density and viscosity of ethanol relative to pure diesel contributes to higher specific fuel consumption.

In the table. 2 are given some of the basic properties of diesel fuel and ethanol.

Table 2. Essential properties of diesel and ethanol

Fuel properties	Unit	Diesel	Ethanol
Chemical formula	-	C_{10}–H_{15}	C_2H_5OH
Density at 20°C	kg/m^3	837	788
Cetane number	-	50	5-8
Viscosity at 40 °C	mm^2/s	2.6	1,2
Surface tension at 20°C	N/m	0,023	0,015
Calorific value	MJ/kg	43	26,8
Specific heat capacity	J/kg°C	1850	2100
Boiling point	°C	180-360	78
Oxygen content	% weight	0	34,8
Latent heat of evaporation	kJ/kg	250	840
Hydrogen content	(wt%)	~14	13
Fuel-air attitude	-	15,0	9,0
Molecular weight	kg	170	46

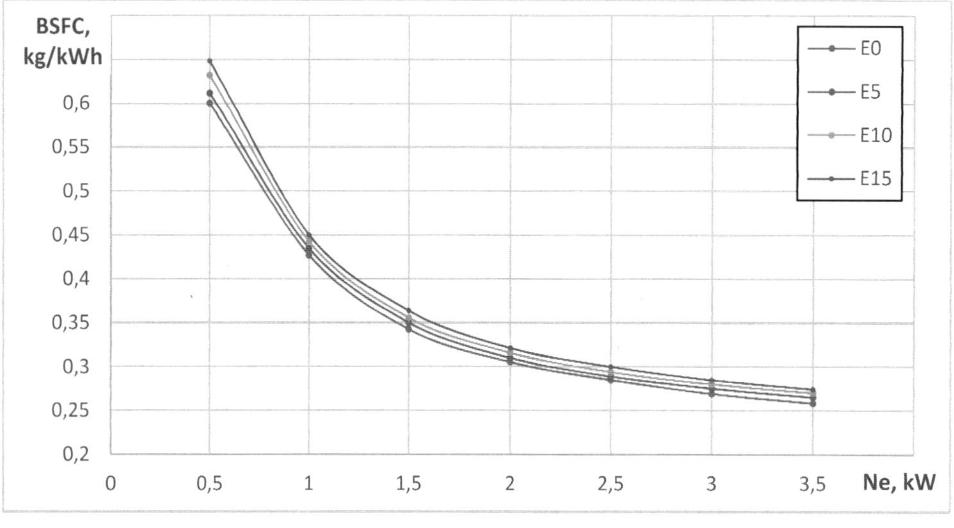

Figure 2. Specific fuel consumption variation at different concentrations of ethanol

Figure 3 shows the variation of the exhaust gas temperature of the engine at a different concentration of ethanol. From the results obtained, it becomes clear that with the increase of the load the temperature of the exhausted gas increases. With an increase in the ethanol content (E10 - E15), a slight increase in the exhaust gas temperature is noticeable and is the highest at E15. This can be explained by the improvement of the combustion process due to the presence of oxygen in the ethanol mixtures. Higher latent heat of ethanol evaporation is the other factor that influences the reduction of exhaust temperature.

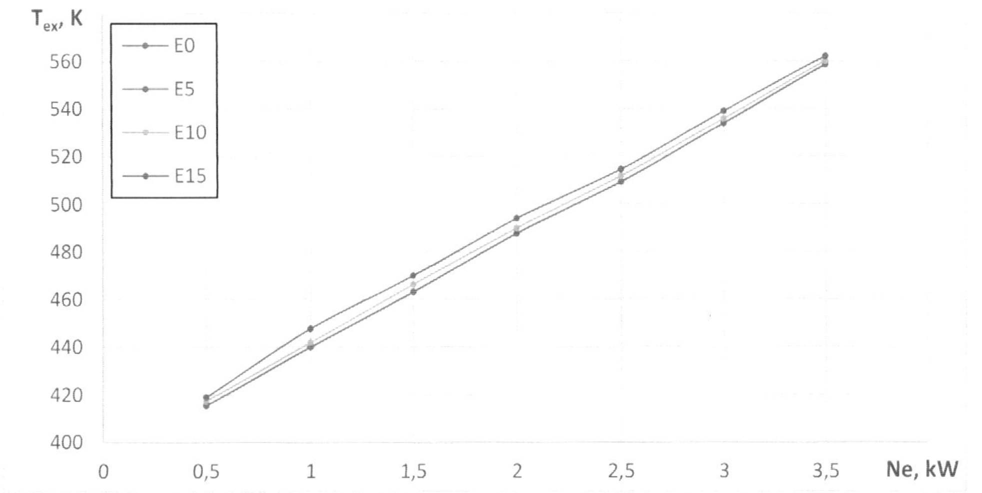

Figure 3. Exhaust gases temperature variation at different ethanol concentrations

The variation of the ignition delay period at different concentrations of ethanol is shown in Fig. 4. By increasing the ethanol content in the mixture (E5 - E15) the cetane number of the mixture is reduced resulting in an ignition delay period. At full load, the ignition delay period of pure diesel fuel is $15{,}62^0$ compared to the E15 mixture, which is $23{,}12^0$. The longer ignition delay period of ethanol mixtures results in the formation of poorer mixtures, which in turn leads to a decrease in the combustion temperature and a reduction in the formation of nitrogen oxides.

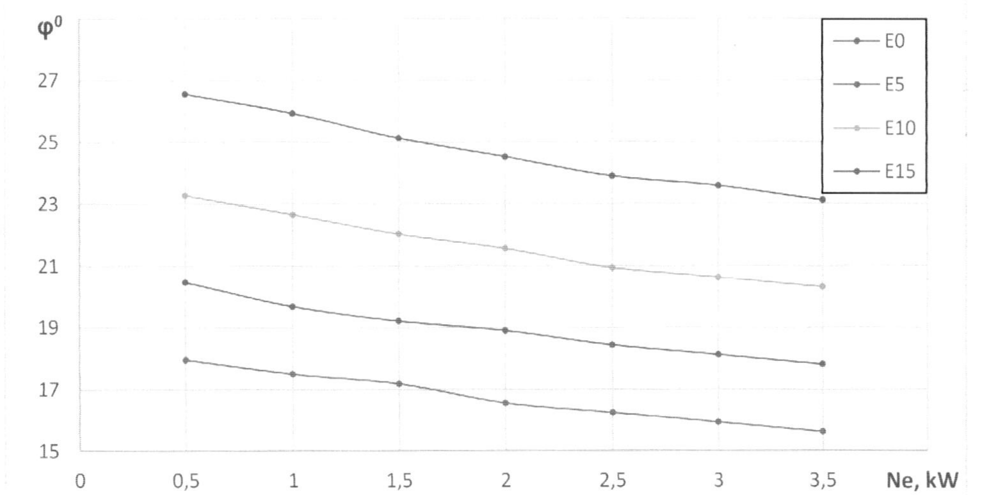

Figure 4. Variation of the ignition delay period at different concentrations of ethanol

The in-cylinder pressure variation, depending on the crankshaft angle, at different ethanol concentrations, is shown in Fig. 5. From the results obtained, it can be seen that the maximum pressure for pure diesel fuel is 7.96 MPa at 50° after TDC, for the mixture E5 is 7.8 MPa at 50° after TDC, for the mixture E10 is 7.56 MPa at 60° after TDC and for E15 is 7.15 MPa at 70° after TDC. Due to the lower cetane number of mixtures with ethanol (E5 - E15) the start of the combustion process occurs later.

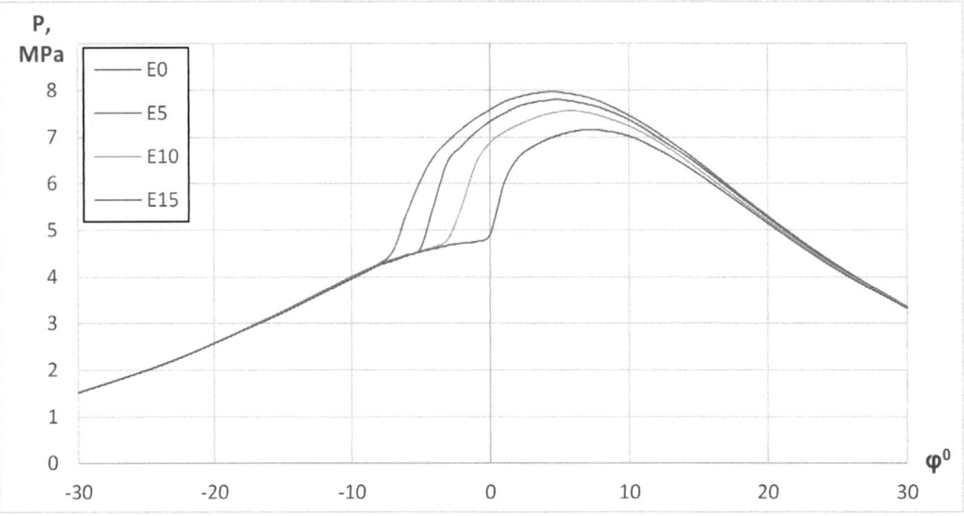

Figure 5. In-cylinder pressure variation depending on the crankshaft angle, with different concentrations of ethanol

Figure 6 shows the variation of NOx emissions depending on the ethanol concentration in the diesel fuel mixture. As the ethanol concentration increases in the mixture, NOx decreases due to the low latent heat of evaporation of the mixture. Another possible cause is the formation of a lean mixture, which leads to a decrease in the combustion temperature.

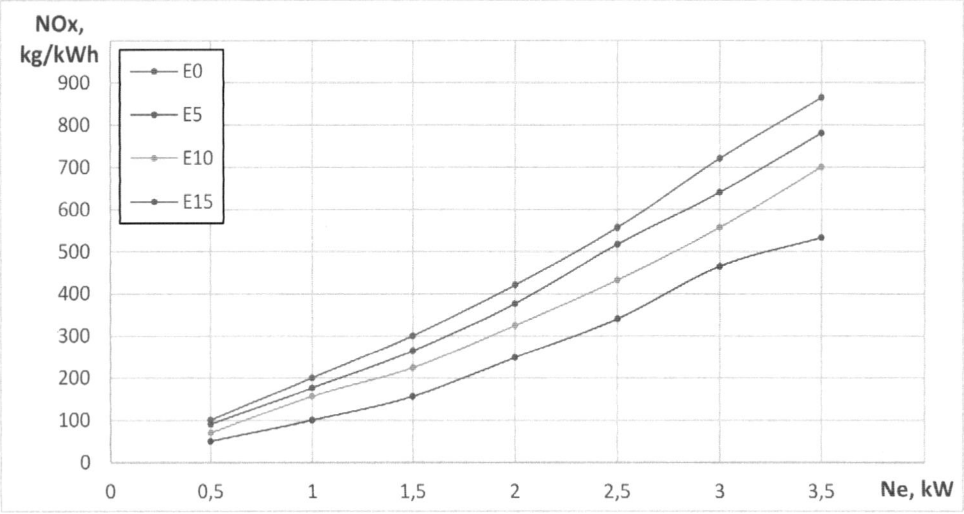

Figure 6. Variation of NOx emissions at different concentrations of ethanol

The CO_2 variation at different concentrations of ethanol is shown in Fig. 7. The figure shows that blending diesel fuel with ethanol results in CO_2 reduction at low loads. Increasing the ethanol content of the mixture reduces the C/H ratio and increases the O_2 content in the cylinder. This circumstance creates a condition for a lean mixture and reduces the temperature at the end of the combustion process. Also, the reduction of CO_2 occurs in parallel with a decrease in the C/H ratio. With low engine load, the combustion temperature is insufficient to initiate full combustion, resulting in an increase in CO_2 emissions for all tested fuels and mixtures. However, at higher loads, combustion temperatures are high enough to achieve fuller combustion, resulting in a reduction in CO_2 for all tested fuels and mixtures. The reduction of CO_2 emissions has also was reported by other researchers [8].

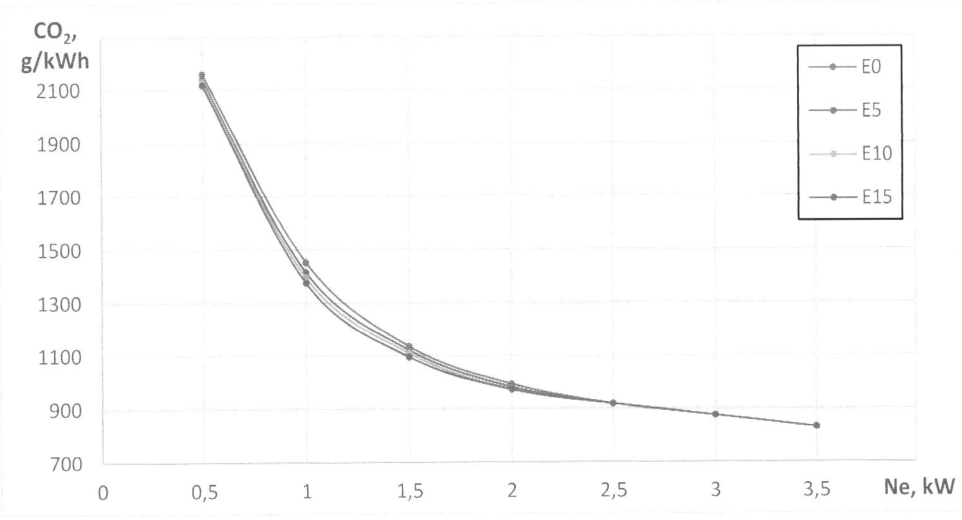

Figure 7. Variation CO_2 emissions at different concentrations of ethanol

3. Conclusion

The main purpose of this study is to investigate the effect of using ethanol and diesel fuel blends on the diesel engine characteristics. According to the simulation results, it can be concluded that:

1. By increasing the ethanol concentration in the mixture (E5-E15), the specific fuel consumption increases, with a maximum specific fuel consumption being obtained in mixtures with 15% ethanol E15.
2. With an increase in the ethanol concentration in the mixture, a slight increase in the exhaust gas temperature is observed, as the highest is at E15.
3. Increasing the ethanol content of the mixture (E5 - E15) reduces its cetane number, resulting in a long ignition delay period. At full load, the ignition delay period for pure diesel fuel is 15,62^0 compared to the E15 mixture, which is 23,12^0.
4. The maximum in-cylinder pressure for pure diesel fuel is 7.96 MPa at 50° after TDC, for the mixture E5 is 7.8 MPa at 50° after TDC, for E10 is 7.56 MPa at 60° after TDC and for E15 is 7.15 MPa at 70° after TDC. Due to the lower cetane number of mixtures with ethanol (E5 - E15) the start of the combustion process occurs later
5. With an increase in the ethanol concentration, the NOx mixture decreases, due to the low latent heat of evaporation of the mixture. Another possible cause is the formation of lean mixture, which leads to a decrease in the combustion temperature.
6. Increasing the ethanol concentration in the mixture results in CO_2 reduction at low loads.

Acknowledgments
The study was supported by contract of University of Ruse "Angel Kanchev", № BG05M2OP001-2.009-0011-C01. Support for the development of human resources for research and innovation at the University of Ruse "Angel Kanchev". The project is funded with support from the Operational Program" Science and Education for Smart Growth 2014 - 2020" financed by the European Social Fund of the European Union.
The present document was supported with the financial assistance of the Project 2019-RU-03.

IOP Conf. Series: Materials Science and Engineering **614** (2019) 012011 doi:10.1088/1757-899X/614/1/012011

References

[1] Iliev S Comparison of Ethanol and Methanol Blending with Gasoline Using Engine Simulation, *Biofuels - Challenges and opportunities*, Mansour Al Qubeissi, IntechOpen, DOI: 10.5772/intechopen.81776.

[2] Iliev S, A Comparison of Ethanol, Methanol and Butanol Blending with Gasoline and Relationship with Engine Performances and Emissions, *Proceedings of the 29th DAAAM International Symposium*, pp.0505-14, B. Katalinic (Ed.), Published by DAAAM International, ISBN 978-3-902734-20-4, ISSN 1726-9679, Vienna, Austria.

[3] Iliev S 2014 Developing of a 1-D combustion model and study of engine performance and exhaust emissions using ethanol-gasoline blends, *IAENG Transaction of Engineering Technologies* (Netherlands, Springer).

[4] Torres-Jimenez E et al. 2011 Physical and chemical properties of ethanol–diesel fuel blends *Fuel*, vol. 90 pp 795-802.

[5] Hansen A 2005 Ethanol-diesel fuel blends. A review *Bioresource Technology* **96** pp. 277-285

[6] Huang Z H et al. 2008 Combustion characteristics of a compression ignition engine fuelled with diesel–ethanol blends *Proceedings of the Institution of Mechanical Engineers, Part D: Journal of Automobile Engineering* vol. 222 pp 265-274.

[7] Kass M D, Thomas J F, Storey J M, Domingo N, Wade J and Kenreck G 2001 Emissions from a 5.9 liter diesel engine fueled with ethanol diesel blends. *SAE Technical Paper* 2001-01-018 (SP-1632).

[8] Ghadikolaei M 2018 Study of combustion, performance and emissions of diesel engine fueled with diesel/biodiesel/alcohol blends having the same oxygen concentration. *Energy* **157** pp 258-269.

IOP Conf. Series: Materials Science and Engineering **614** (2019) 012012 doi:10.1088/1757-899X/614/1/012012

Modeling and investigation of a diesel engine with methanol additives

S Iliev*, E Mitev

University of Ruse, Department "Engines & vehicles", 8 Studentska Str., Ruse, Bulgaria

*E-mail: spi@uni-ruse.bg

Abstract. In this study, the effects of methanol–diesel (M5, M10 and M15) fuel blends on the emissions of a diesel engine were investigated. In the simulation study, the single-cylinder, four-stroke, direct injection compression ignition engine was used. The results obtained from the use of methanol–diesel fuel blends were compared to those of pure diesel fuel (M0). One-dimensional simulation of a single-cylinder diesel engine with the help of the GT-Power software has been done. This software is capable of predicting the exhaust emissions. The simulation is designed to test the characteristics of the diesel engine working with methanol-diesel blends at full load.

1. Introduction

Alternative fuels are derived from resources other than oil. Using these fuels in internal combustion engines, they emit less harmful emissions than petroleum fuels and most of them are more cost-effective than oil. Also, they are renewable. The most commonly used fuels as alternative fuels are natural gas, propane, ethanol, methanol and hydrogen [1]. The idea of adding small amounts of ethanol or methanol to diesel is not new, it dates back to the 1970s when oil supplies were limited and demand for alternative energy sources began to replace the use of diesel fuel [2]. Alternative fuels are gaining more and more attention, as emissions they of pollutants from internal combustion engines, higher fuel costs and depletion of crude oil [3]. Alcohols, such as ethanol and methanol, are the subject of much research on the part of many scientists, because they are oxygen-containing. Alcohols produced from bio-resources are widely used in compression-ignition engines as additional fuel or as a mixture with the main diesel fuel. However, methanol has an advantage due to its low cost and higher oxygen content [4].

2. Exposition

The study of the literature shows that a number of experimental studies have been carried out to determine the environmental, performance and economic characteristics of compression ignition engines using diesel and methanol mixtures. In one of the studies [5], the author studied a four-stroke, atmospheric, single-cylinder diesel engine with direct injection of a mixture of diesel fuel and methanol. The researcher has observed CO, HC and soot emissions reduction, but NOx emissions raise with increasing methanol concentration in the mixture. The author also observed an increase in fuel consumption. In another study [6], it was observed that the addition of methanol to diesel fuel reduces CO, HC and soot emissions due to the oxygen content of methanol. On the other hand, the oxygen content has resulted in improved combustion, which has increased CO_2 emissions. Emissions of NOx have also increased due to the higher rise in temperature during combustion. The low calorific value of methanol has also led to increased fuel consumption. The same trend was observed in two other different studies [7, 8]. But experimental research is not always feasible in terms of time, human strength, and cost. Thus, numerical simulations using appropriate mathematical models to assess the effect of using methanol in diesel engines may be an economical approach to designing a new engine. When designing a specific engine, its design parameters can be easily changed to give an optimal model.

To build the one-dimensional engine model, 1D simulation code of the GT-Power software was used. The software is a leading modeling tool for internal combustion engines used by almost all engine and vehicle manufacturers and is suitable for analysis of a wide range of problems. For engine modeling, specific engine geometry information such as piston diameter and stroke, compression ratio and power, working volume, etc. are required. The characteristics of the engine are given in table. 1, and the model developed in Fig 1.

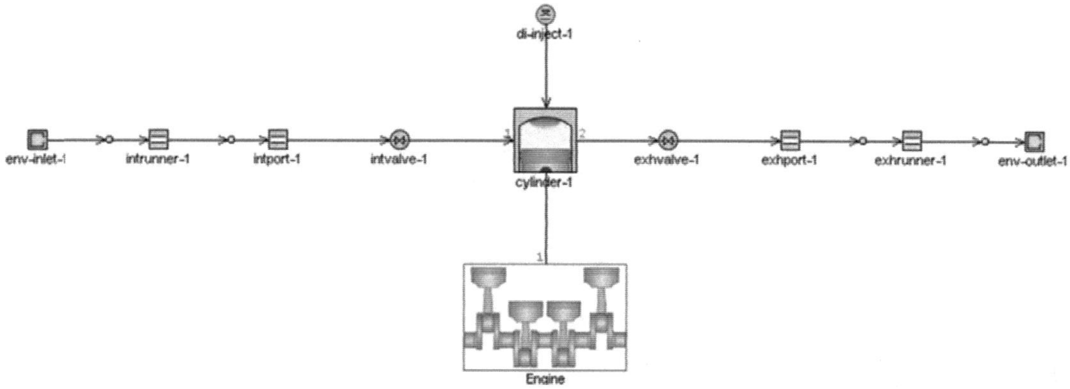

Figure 1. Single-cylinder diesel engine model.

The characteristics of the diesel engine used in the study are shown in table 1.

Table 1. Characteristics of the diesel engine.

Description	Value
Fuel	Diesel
Engine type	Single cylinder
Piston diameter [mm]	87,5
Piston stroke [mm]	110
Volume [cc]	661
Compression ratio	17,5:1
Power [kW]	3,5
Maximum rotation speed [min^{-1}]	1500
Opening suction valve before TDC [^{0}]	4,5
Closing suction valve after BDC [^{0}]	35,5
Opening exhaust valve before TDC [^{0}]	35,5
Closing exhaust valve after BDC [^{0}]	4,5

Figure 1 shows a single-cylinder diesel engine developed using the GT-Power software. Once the engine model has been developed, the engine's fuel has to be defined. For this purpose, different thermophysical properties of diesel and methanol mixtures (5%, 10% and 15%) were introduced into the software as input data. This study was conducted at an engine speed of n = 1500 min^{-1}, a compression ratio of 17.5: 1 and an injection timing of 23° before TDC.

In the table 2 are given some of the basic properties of diesel fuel and methanol.

Table 2. Essential properties of diesel and methanol.

Fuel properties	Unit	Diesel	Methanol
Chemical formula	-	C_{10}~H_{15}	CH_3OH
Density at 20°C	kg/m^3	837	788
Cetane number	-	50	5-8
Viscosity at 40 °C	mm^2/s	2.6	1,2
Surface tension at 20°C	N/m	0,023	0,015
Calorific value	MJ/kg	43	26,8
Specific heat capacity	J/kg°C	1850	2100
Boiling point	°C	180-360	78
Oxygen content	% weight	0	34,8
Latent heat of evaporation	kJ/kg	250	840
Hydrogen content	(wt%)	~14	13
Fuel-air attitude	-	15,0	9,0
Molecular weight	kg	170	46

Figure 2 shows the variation of the specific fuel consumption at different methanol concentrations of 0% Methanol (M0), 5% Methanol (M5), 10% Methanol (M10), 15% Methanol. The study was performed at an engine speed of 1500 min-1. The results obtained show that increasing the methanol concentration in the mixture, (M5 - M15), the specific fuel consumption increases, with an increase in the specific fuel consumption for M5 of 2.59%, M10 4.53% M15 is 6.16%. The lower methanol calorific value of 19.7 compared to pure diesel fuel 43 (Table 1) increases the fuel consumption to obtain the same amount of energy. Also, the low density and viscosity of methanol relative to pure diesel contributes to higher specific fuel consumption.

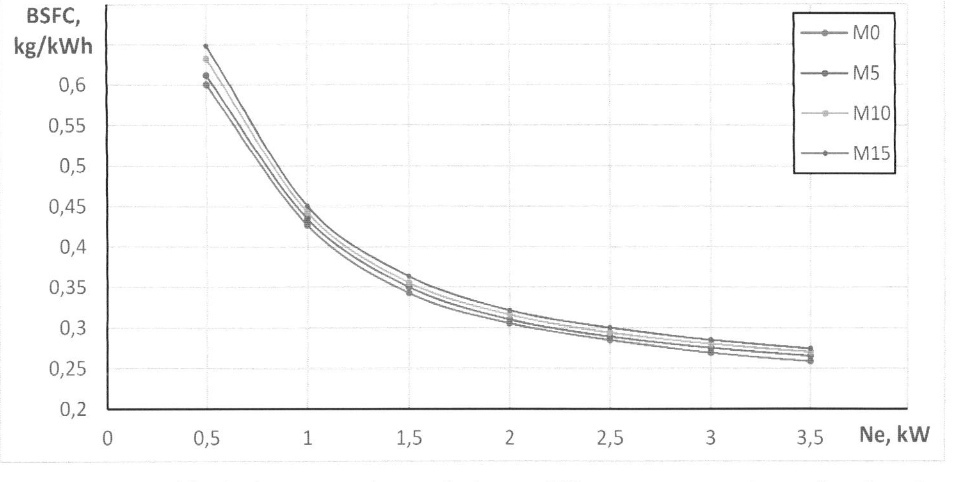

Figure 2. Specific fuel consumption variation at different concentrations of methanol.

Figure 3 shows the variation of the exhaust gas temperature of the engine at a different concentration of methanol. From the results obtained, it becomes clear that with the increase of the load the temperature of the exhausted gas increases. With an increase in the ethanol content (M10 - M15), a slight increase in the exhaust gas temperature is noticeable and is the highest at M15. This can be explained by the improvement of the combustion process due to the presence of oxygen in the ethanol mixtures. Higher latent heat of methanol evaporation is the other factor that influences the reduction of exhaust temperature.

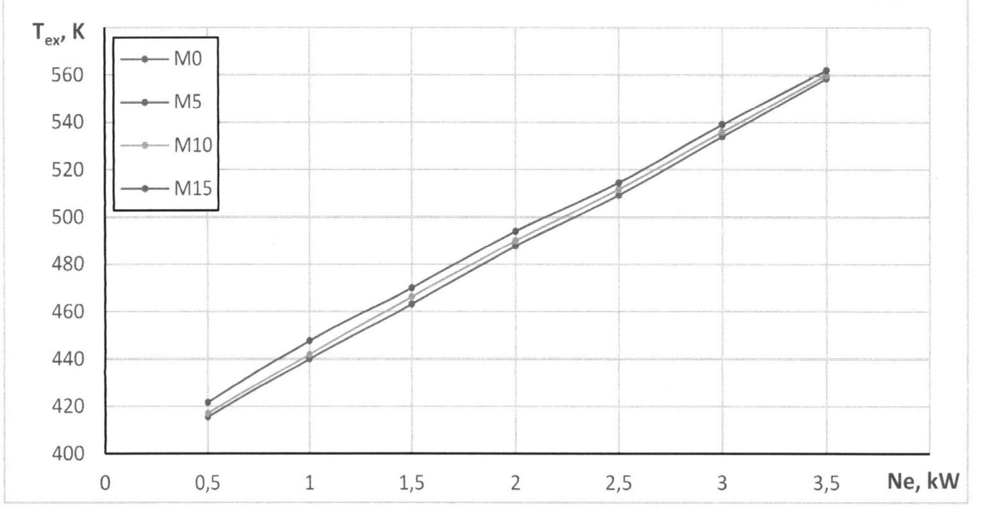

Figure 3. Exhaust gases temperature variation at different methanol concentrations.

The variation of the ignition delay period at different concentrations of methanol is shown in Fig. 4. By increasing the methanol content in the mixture (M5 - M15) the cetane number of the mixture is reduced resulting in a long ignition delay period. At full load, the ignition delay period of pure diesel fuel is 15,620 compared to the M15 mixture, which is 23,120. Longer ignition delay period of methanol mixtures results in the formation of poorer mixtures, which in turn leads to a decrease in the combustion temperature and a reduction in the formation of nitrogen oxides.

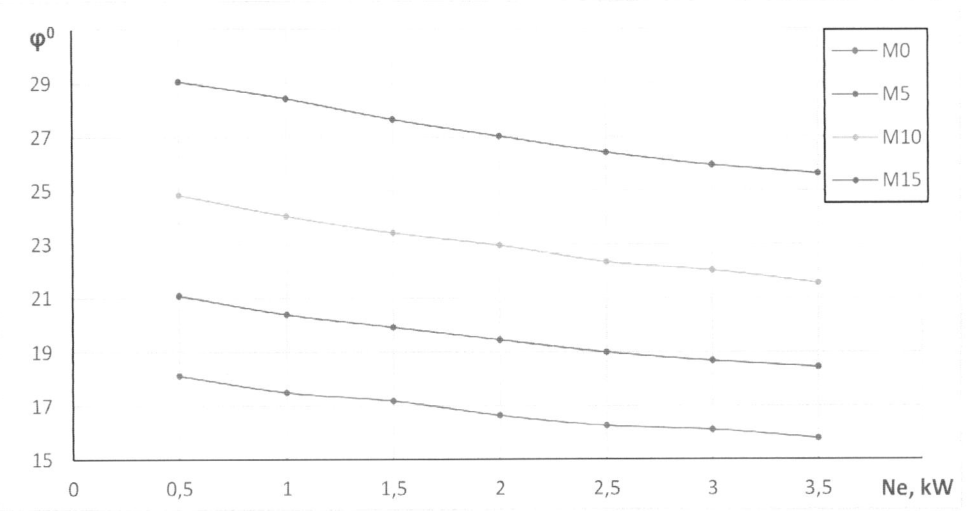

Figure 4. Variation of the ignition delay period at different concentrations of methanol.

The in-cylinder pressure variation, depending on the crankshaft angle, at different methanol concentrations, is shown in Fig. 5. From the results obtained, it can be seen that the maximum in-cylinder pressure for pure diesel fuel is 7.96 MPa at 5°C after TDC, for the mixture M5 is 7.7 MPa at 8°C after TDC, for the mixture M10 is 7.38 MPa at 6°C after TDC and for M15 is 6.72 MPa at 6°C after TDC. Due to the lower cetane number of mixtures with methanol (M5 - M15) the start of the combustion process occurs later.

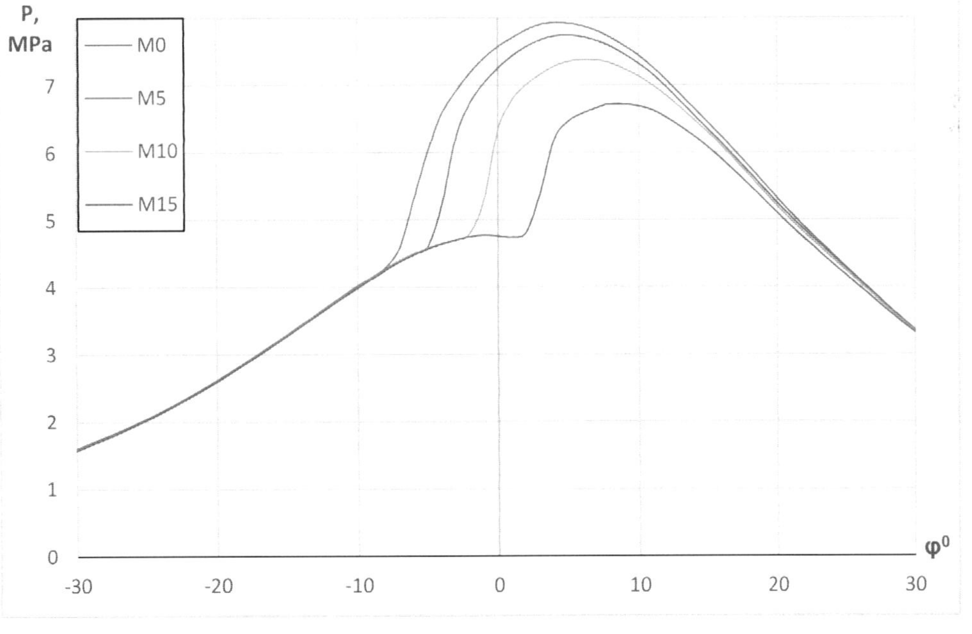

Figure 5. In-cylinder pressure variation depending on the crankshaft angle, with different concentrations of methanol.

Figure 6 shows the variation of NOx depending on the methanol concentration in the diesel fuel mixture. As the methanol concentration increases in the mixture, NOx decreases due to the low latent heat of evaporation of the mixture. Another possible cause is the formation of a lean mixture, which leads to a decrease in the combustion temperature.

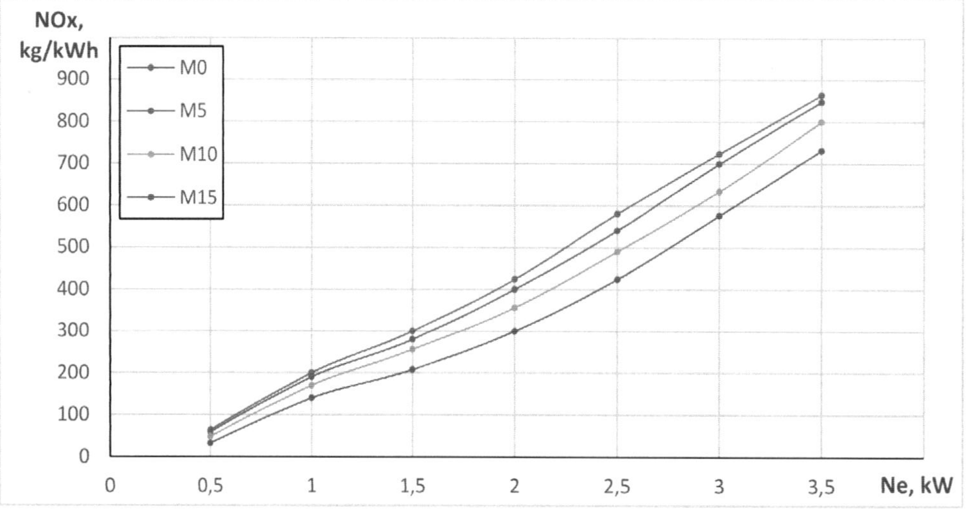

Figure 6. Variation of NOx emissions at different concentrations of methanol.

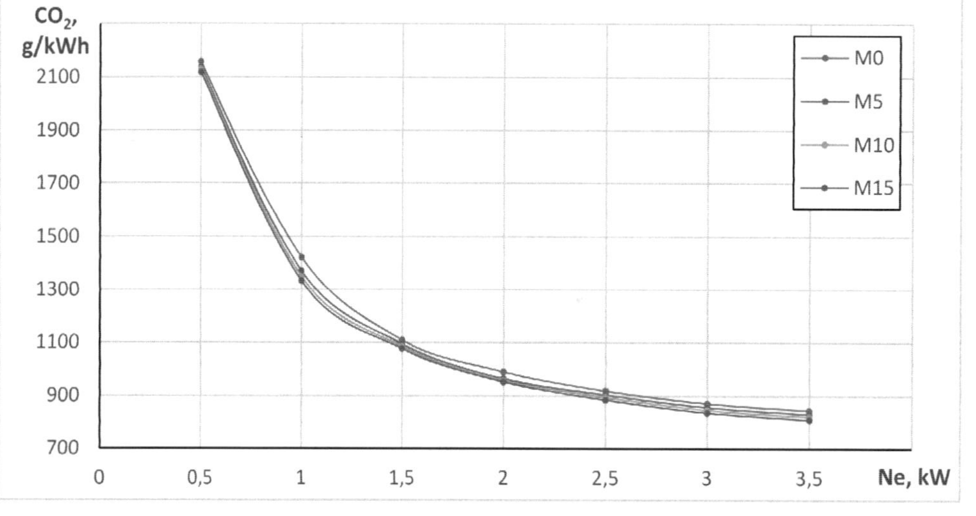

Figure 7. Variation CO_2 emissions at different concentrations of methanol.

The CO_2 variation at different concentrations of methanol is shown in Fig. 7. The figure shows that blending diesel fuel with methanol results in an inessential reduction in CO_2. Increasing the methanol content of the mixture reduces the C/H ratio and increases the O_2 content in the cylinder. This circumstance creates a condition for a lean mixture and reduces the temperature at the end of the combustion process. Also, the reduction of CO_2 occurs in parallel with a decrease in the C/H ratio. With low engine load, the combustion temperature is insufficient to initiate full combustion, resulting in an increase in CO_2 emissions for all tested fuels and mixtures. However, at higher loads, combustion temperatures are high enough to achieve fuller combustion, resulting in a reduction in CO_2 for all tested fuels and mixtures. The reduction of CO_2 emissions has also been established by other researchers.

3. Conclusion

The main purpose of this study is to investigate the effect of using methanol and diesel fuel blends on the diesel engine characteristics. According to the simulation results, it can be concluded that:

1. Increased methanol concentration in the mixture (M5-M15), the specific fuel consumption increases, with the increase in specific fuel consumption for M5 being 2.59%, for M10 4.53% and for M15 6.16%.
2. With an increase in the methanol concentration in the mixture, a slight increase in the exhaust gas temperature is observed, as the highest is at M15.
3. Increasing the methanol content of the mixture (M5 - M15) reduces its cetane number, resulting in a long ignition delay period. At full load, the ignition delay period for pure diesel fuel is 15,620 compared to the M15 mixture, which is 23,120.
4. The maximum in-cylinder pressure for pure diesel fuel is 7.96 MPa at 50° after TDC, for the mixture M5 is 7.7 MPa at 80° after TDC, for M10 is 7.38 MPa at 60° after TDC and for M15 is 6.72 MPa at 80° after TDC. Due to the lower cetane number of mixtures with methanol (M5 - M15) the start of the combustion process occurs later
5. With an increase in the methanol concentration, the NOx mixture decreases, due to the low latent heat of evaporation of the mixture. Another possible cause is the formation of lean mixture, which leads to a decrease in the combustion temperature.
6. Increasing the methanol concentration in the mixture results in an inessential reduction in CO_2.

Acknowledgments
The study was supported by contract of University of Ruse "Angel Kanchev", № BG05M2OP001-2.009-0011-C01. Support for the development of human resources for research and innovation at the University of Ruse "Angel Kanchev". The project is funded with support from the Operational Program" Science and Education for Smart Growth 2014 - 2020" financed by the European Social Fund of the European Union.
The present document was supported with the financial assistance of the Project 2019-RU-03.

References

[1] Iliev S Comparison of Ethanol and Methanol Blending with Gasoline Using Engine Simulation, Biofuels - Challenges and opportunities, Mansour Al Qubeissi, IntechOpen, DOI: 10.5772/intechopen.81776.
[2] Iliev S, A Comparison of Ethanol, Methanol and Butanol Blending with Gasoline and Relationship with Engine Performances and Emissions, *Proceedings of the 29th DAAAM International Symposium* pp.0505-14 B. Katalinic (Ed.) (DAAAM International)
[3] Iliev. S, Hadjiev K 2014 Theoretical Study of Engine Performance Working on Methanol-Gasoline Blends *Proceedings of the World Congress on Engineering* Vol II, WCE 2014, (London, U.K).
[4] Torres-Jimenez E et al. 2011 Physical and chemical properties of ethanol–diesel fuel blends *Fuel*, vol. 90 pp 795-802.
[5] Sayin C 2010 Engine performance and exhaust gas emissions of methanol and ethanol–diesel blends *Fuel* **89** pp. 3410–15.
[6] Sayin C An experimental investigation on the effect of injection pressure on the exhaust emissions of a diesel engine fueled with methanol-diesel blends *Energy Sources*, Part A, 33, 2011, pp. 2206–2217.
[7] Sayin C, Ozsezen A N and Canakci M 2010 he influence of operating parameters on the performance and emissions of a DI diesel engine using methanol-blended-diesel fuel, *Fuel*, **89** pp. 1407–14.
[8] Sayin C, Ilhan M, Canakci M and Gumus M 2009 Effect of injection timing on the exhaust emissions of a diesel engine using diesel–methanol blends *Renewable Energy* **34** pp. 1261–69.

IOP Conf. Series: Materials Science and Engineering **614** (2019) 012013 doi:10.1088/1757-899X/614/1/012013

Analysis of the Capabilities of Pistons Design using SOLIDWORKS models

Delyan Petkov

Department of Automotive Engineering and Technologies, Technical University - Varna, 1 Studentska Str., city of Varna, Bulgaria

E-mail: delyan.petkov@tu-varna.bg

Abstract. The difficulty in designing internal combustion piston is determined by complex operating conditions. Piston is the busiest engine part. Combined mechanical and thermal loads work on it. Profiling the piston in two mutually perpendicular planes is a major part of its design. The shape of the piston in a cold state is quite different from the cylindrical one. The aim is, despite the varying amount of material in the various zones, when the working temperature reaches the piston to take cylindrical form, thus providing minimal clearance to the engine cylinder. Because it is a moving part, the parameters at work are hard to determine. There are studies of the piston temperature state, some of which are described and analysed.

1. Introduction

The work presented by me aims to analyse the possibilities of widely accessible SOLIDWORKS software for designing pistons for modern diesel engines.

Profiling the piston in two mutually perpendicular planes is a major part of the piston design. By determining the method to set boundary conditions in SOLIDWORKS simulation, I receive the piston temperature field, and the stresses from thermal and mechanical loads, and compare the resulting displacements with the dimensions of a real piston in cold state. The temperatures at specific piston points obtained experimentally [1] are criteria for the result adequacy.

Given the fact that piston is a moving part, the parameters at work are hard to determine. There are studies of the piston temperature state, some of which are described and analysed in [2].

The subtask of the work I present with the publication is to determine the method to set boundary conditions and to assess the adequacy of the results obtained.

2. Explanation

Creation of Three-dimensional Piston Image

We create three-dimensional images based on the real part. The design dimensions are taken from the basic dimensions specified in the manufacturer's technical documentation.

When drawing the piston, its microgeometry is not taken into account. The created model has cylindrical shape. Figure 1 shows a diagram of the dimensions of a piston of Perkins

IOP Conf. Series: Materials Science and Engineering **614** (2019) 012013 doi:10.1088/1757-899X/614/1/012013

Prima engine, which is subject to the study The diagram is taken from the engine technical documentation.

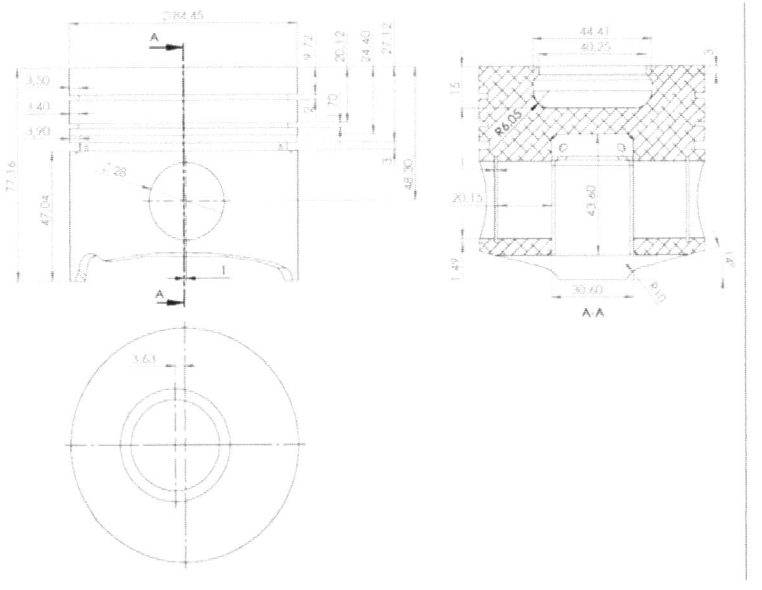

Figure 1. Piston diagram

The parts are dimensioned in accordance with English technical documentation, where the measurement system is in inches, and therefore the dimensions in millimetres have inaccurate values. Since the objective is to recreate the real parts in drawing, rounding has not been made.

Figure 2 shows the finished piston model.

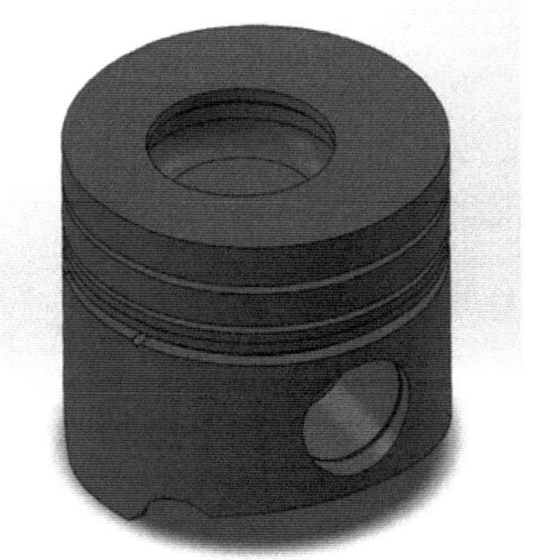

Figure 2. Piston model.

2.1. Boundary Conditions

a. Attachment

IOP Conf. Series: Materials Science and Engineering **614** (2019) 012013 doi:10.1088/1757-899X/614/1/012013

The manner of attaching the piston that determines its contact with the cylinder and the connection with rod via the piston bolt is detailed in [3].

b. Load

In order to evaluate the impact of the different types of load on the total stress and strain state of the piston two simulations are made. In the first simulation, only mechanical load caused by gas pressure and inertial force is applied to the piston.

To analyse the piston temperature field, separate test is conducted in the Thermal option. These are third generation boundary conditions. The values of heat transfer coefficient αg.mid. [W / m^{2}deg] and the resulting temperature tr.res [^{o}C] are entered in the option.

Table 1. Thermal boundary conditions.

Zone	$A_{g.mid.}$	$T_{g.res.}$
2	446	86
3	598	95
5	748	109
6	600	109
7	388	109
12	26148	186
13	22	186
14	26,148	186
15	744	179
16	34241	196
17	29	200
18	34,241	200
19	73	560
20	137	991
21	271	1,559
22	374	1,559

IOP Conf. Series: Materials Science and Engineering **614** (2019) 012013 doi:10.1088/1757-899X/614/1/012013

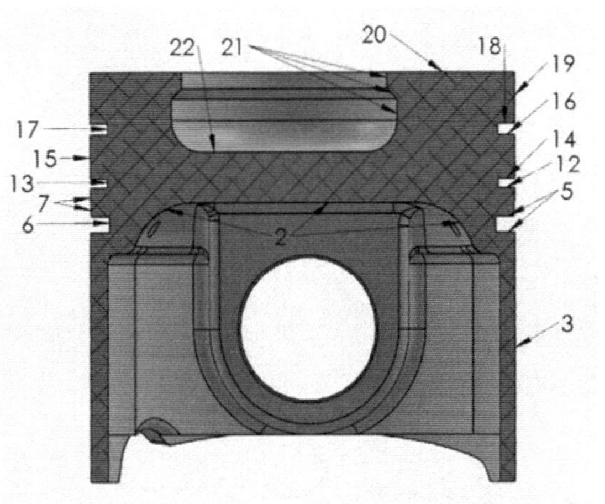

Figure 3. Thermal load setting zones.

The values of $\alpha_{g.mid}$ and $t_{g.res}$ for the different zones are determined theoretically and experimentally. As regards the current tests, the values obtained for diesel engine in [1] have been taken into account. They are shown in the table 1.

The second simulation combines mechanical and thermal deformations with the data obtained for the temperature field.

In SOLIDWORKS environment, the piston is subjected to mechanical load in the Static option, using the Pressure and Gravity commands.

The method of conducting mechanical, thermal and combined load simulations and the resultant thermal field are detailed in a previous study [2].

The deformed state of the piston resulting from combined action of mechanical and thermal load is illustrated on figure 4. The maximum displacement is 0.348 mm.

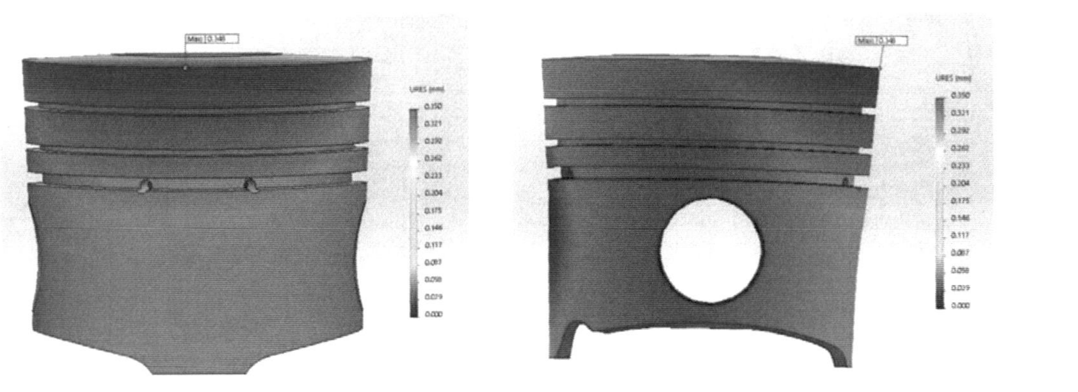

Figure 4. Deformations

The results of deformation of the piston and displacements of points from its surface in running order are the starting point for piston profiling.

Radial displacements by piston height are determined in two mutually perpendicular planes (along the axis of piston bolt and perpendicular to the axis of piston bolt); the profile of piston longitudinal generant must be the mirror curve of the diagrams shown in figure 5, i. e. the piston expansion in a given zone during operation must be equal to its shrinking in a cold state.

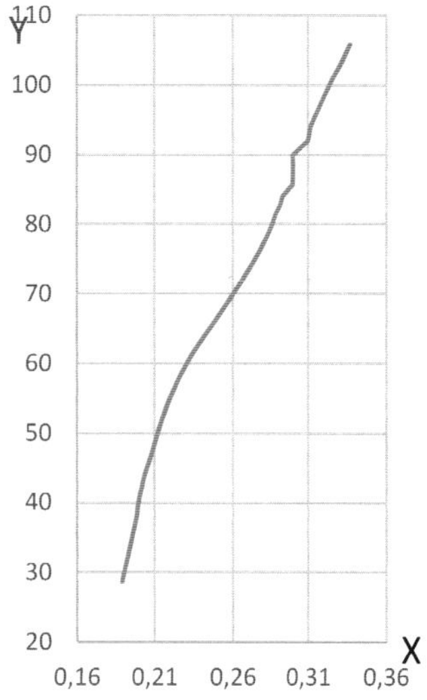

a.

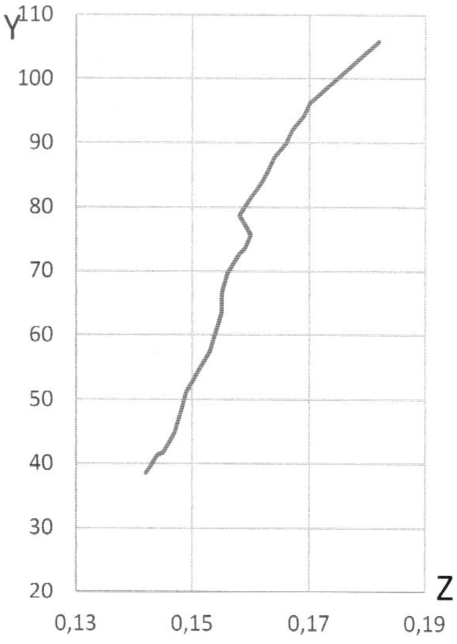

b.

Figure 5. Diagrams of radial thermal expansions

a - perpendicular to the axis of piston bolt; b - along the axis of piston bolt

2.2. Real Piston Microgeometry Measurement

Piston of the same engine is measured with micrometre. The results are shown in Table 2, with Y-axis levels being the same as those indicated in the graphic of SOLIDWORKS report for convenience during comparison.

Table 2. Piston dimensions

Y	**X [mm]**	**Z [mm]**
100	84.12	84.27
90	84.15	84.28
85	84.16	84.29
80	84.17	84.29
60	Piston bolt hole	84.3
40	84.27	84.31

3. Conclusion.

The analysis shows that the approach selected in determining the boundary conditions for attachment, mechanical and thermal load is appropriate.

The created model provides adequate results for the piston temperature expansion.

This model can be used to determine the piston shape at the initial stage of its design.

The method can be used in the training of students in the subject Computer Methods to design Internal Combustion Engines of the Master programme of Department of Transport Engineering and Technologies of Technical University - city of Varna.

4. References

[1] Serafimov M 2000 The thermal boundary conditions of the piston of an automotive diesel engine as a function of the working mode, Sofia, *J. Motauto 2000* 365.

[2] Belchev S and Petkov D Evalution of the possibilities for modeling the heat and deformation condition of an engine piston in SOLIDWORKS *J. EcoVarna* **24** 367.

[3] Petkov D and Belchev S Capability of the finite element method in SOLIDWORKS envoirment for determining the actual thermal and deformation state of a diesel engine piston *J. 56th Science Conference of Ruse University Bulgaria 2017* **v56 book4** 420.

[4] Karavasilev O, Kozhuharov M, Dimitrov N, Grigorov B and Mitrev R 2012 *SolidWorks-Basic modeling and drawings* (Sofia: TechnoLogica) p 600.

Acknowledgments

The report was prepared with the help of 2019 Young Scientist and Postdoctoral Fellow Programme.

IOP Conf. Series: Materials Science and Engineering **614** (2019) 012014 doi:10.1088/1757-899X/614/1/012014

Hyperspectral environmental monitoring

N D Zdravcheva

University of Architecture, Civil Engineering and Geodesy, bul. "Hristo Smirnenski" 1, Sofia, Bulgaria

E-mail: neli_z @ abv.bg

Abstract. The report has made a critical analysis of modern hyper-spectral methods and their importance for the realization of reliable and effective control of the state of natural resources. The spectral characteristics of some objects and the creation of spectral libraries are examined. Attention is paid to the principles, ways and means of obtaining the hyper-spectral images, their specifics and characteristics, the trends in their development and their use in solving different socio-economic and scientific-applied tasks in the field of environmental monitoring. It also analyses its organization with a view to the effective use of complex data supplied by space, aviation and ground-based sources.

1. Introduction

An important role at reaching ecological and stable growth of our society plays providing operational information about real conditions of natural objects and their ecological characteristics. From one side, it should be with proved quantities (authenticities, clearness, actuality and others), from other side are very important the ways for its preservations, availability and working tools. From [2] the author analyzes the important role and meaning of the symbioses between photogrammetry and remote sensing and geographical informational systems (GIS). Hyperspectral ecological monitoring appears as one new innovative aspect from contemporary remote sensing techniques. Over the last years to receive better and more quantitative information about characteristics and the real physics-chemical condition of the studied parts of the earth surface and placed above her lots of natural and anthropogenic features, hyperspectral images are used.

2. Hyperspectral methods briefly

The base for hyperspectral methods in remote sensing is so called spectral analysis. It is based on natural ability of every material object to absorb and affect in different ways electromagnetic energy. The application of spectral data analysis, delivered from remote sensing, are based on the assumption that exists compliance between reflected signal (intensity and polarization) and physics-chemical composition of object's reflective surface. Hyperspectral analysis means that spectral analysis that is performed with using hyperspectral images. They consist of large amount of images, made at the same moment above the same area from the earth surface, but at different and many spectral diapasons (parts from the electromagnetic spectrum with different wavelength), from which is registered the intensity of reflected signals from objects. It can be generalized that main dimensions which are registered and processed during hyperspectral analysis are wavelength, intensity of reflected signal

IOP Conf. Series: Materials Science and Engineering **614** (2019) 012014 doi:10.1088/1757-899X/614/1/012014

and spatial coordinates of studied objects. In order to obtain visual idea and easier spatial perception of hyperspectral data analysis can be used the comparison with hypercube which means tridimensional data array - figure 1. This cube contains layers (hundreds and thousands - the number of spectral diapasons from which is registered reflected signal), i.e. from separated two dimensional images, containing from lots of pixels, whose values matches the intensity of reflected signal from every single spectral diapason. Said in other words, data hypercube contains spatial information of the objects and additional spectral information for each pixel from every image. This is why hyperspectral images represents tridimensional pictures.

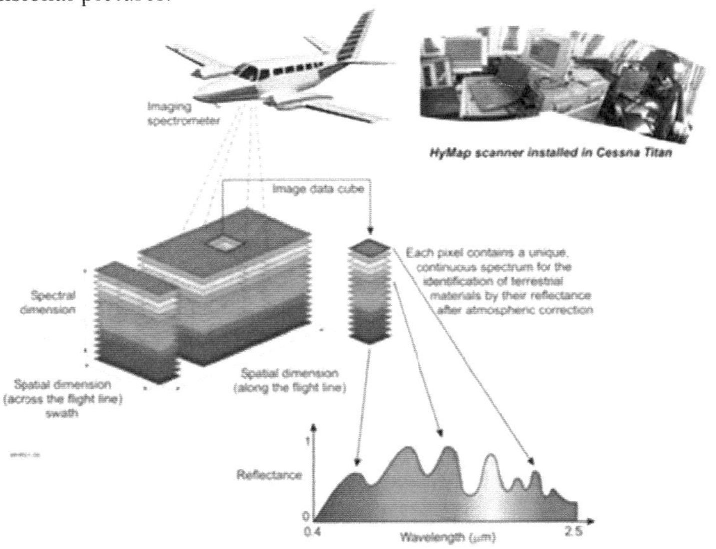

Figure 1. Interpretation of hypercube.

Hyperspectral images can be produced using hyperspectrometres. They are tools that can make at the same time measuring and registration of the spatial coordinates of the objects and the intensity from the reflected signals from them for each of the spectral diapasons with defined wavelengths. The technology of hyperspectral images can be seen on figure 2 [1].

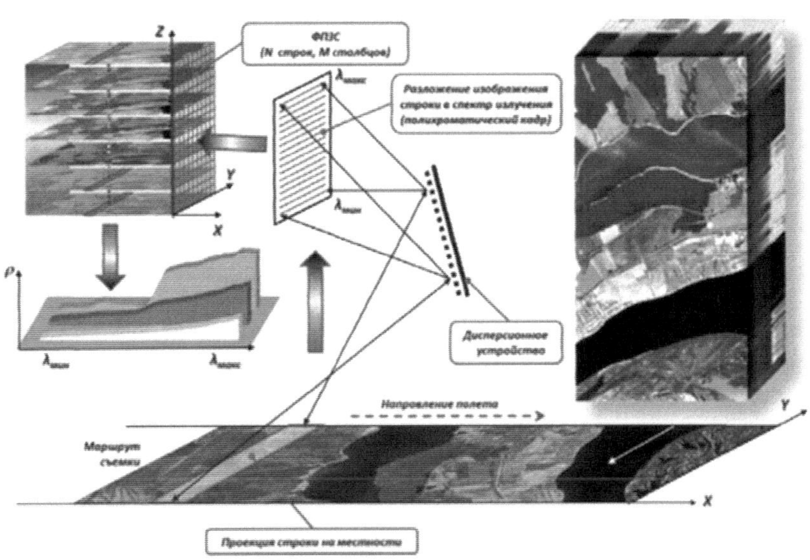

Figure 2. Hyperspectral imagery.

In order to reveal the benefits of hyperspectral analysis is made brief comparison between it and multispectral analysis. Main image characteristics, used in remote sensing is their spectral resolution. It can be determined from the number and width of spectral diapasons of the images. The width is measured with nanometres or micrometres. The lowest spectral resolution (hundreds nanometres) have the photographic panchromatic images and images, received from infrared radiometers, and the highest (till ten nanometres) - hyperspectral images. Said in other words, spectral resolution can be defined from characteristics of these intervals (with defined wavelengths) from electromagnetic spectrum. Images with high spectral resolution provide detailed and rich information and allow accurate analysis and classification from different earth's objects. For instance, from hyperspectral images can be produced full spectral characteristics and data about absorbing water bodies, chlorophyll, minerals, gases at the atmosphere, anthropogenic objects and etc. Exactly they are the main objects for ecological monitoring researches.

Multispectral images contain from series of images about one territory, made at the same time with some (usually ten) different spectral diapasons, which are chosen to be informative for the studied object or easy to registrate. Usually they are produced with using color filters. On the next figure 3 can be seen multispectral images, made from Landsat. Under every image are shown the numbers of layers and the corresponding diapason from the electromagnetic spectrum.

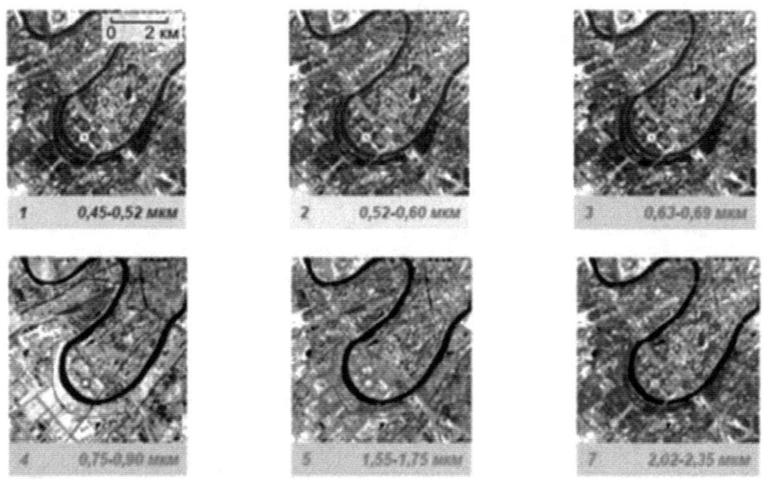

Figure 3. Different multispectral images.

The distinction between hyperspectral and multispectral images can be made from the number of spectral diapasons (bands) or by the type of measurements. From the multispectral method the images are received discreetly and series of single images are made. Said in other words, multispectral images are received from the principe of color photography. Multispectral sensor can have lots of bands, covering the spectre from visible to long wavelength infrared field. Multispectral images can not reproduce the whole spectrum of the object. Whereas hyperspectral images have information about hundreds or thousands spectral diapasons. With hyperspectral analysis is made one principal and fundamental improvement of the data information. The uniqueness of the hyperspectral methods, used in remote sensing is expressed at the ability of specialized sensors to provide receiving of the large amount of images, above which is registered electromagnetic reflection in hundreds and thousands spectral diapasons [6]. Said in other words, hyperspectral imaging produces multidimensional image. As said above, its two dimensions defines the spatial position of the concrete point from the earth surface on the image, and the third -its spectral characteristics in spectral diapason. I.e. for every specific elementary area form the image (pixel) are storage spectral characteristics of its equivalent from the earth surface. Exactly based on them are made remotely physics-chemical properties of the studied natural and anthropogenic objects. For example, hyperspectral methods give a chance not only to define single trees species but also something very important - to held classification inside every vegetation class and to define many biological and other characteristics,

At [3] the author consider different vegetation indices, at [4] their applications. It must be noticed that vegetation indices lay at the basis of the smart and intelligent agriculture. Before the invention of recording hyperspectral techniques for the objects in remote sensing, for receiving object's information are used only multispectral images, which appears set of photographers, made with the help of different color filters.

Hyperspectral method, which provide spatial picture of the spectral distribution, firstly is made and used only for military purposes. Later began the development its mass and effective application in photogrammetry and remote sensing. Nowadays, hyperspectral method is used at other areas as bio-medical optics. In near future can be expected hyperspectral method to become one of the basis technologies in medicine.

3. Spectral characteristics of natural objects

At remote sensing is used primarily information from reflected energy, which is registrated mainly on different aerial and cosmic images. In so-called discriminant analysis of natural objects with spectral reflective characteristics is made studying of the earth surface with using the ability of different surfaces and objects to reflect in different ways the falling electromagnetic energy. Although the high diversity of objects in different landscape, they can be divided in four classes, each of them is characterized with spectral brightness curve. On figure 4 are shown typical curves of spectral brightness for each of them. The following labelling are used: 1- snow, 2-vegetation, 3- soil, 4-water.

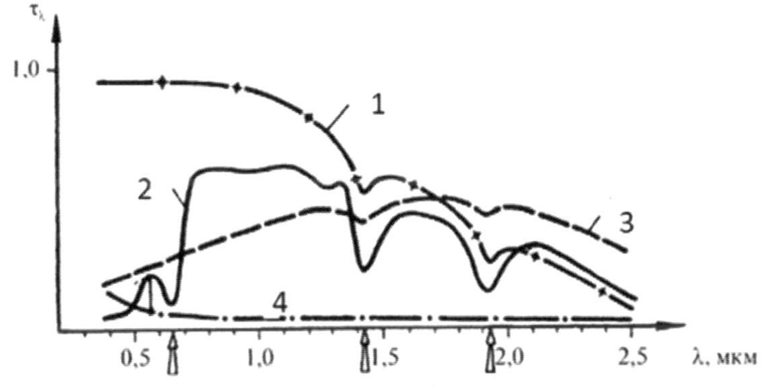

Figure 4. Typical curves of spectral brightness.

1 class - covering snow surfaces and near clouds, which have the highest values of spectral brightness with small decrease in near infrared area. The decrease sharply increases with water saturation.

2 class - it contains vegetation cover, which distinguishes with its typical maximum of reflected ability and very sharply increase in near infrared area.

3 class - it covers soils and rocks and can be characterized with increasing of the spectral brightness when approaching near red area of the electromagnetic spectrum. On figure 5 can be seen spectral characteristics of some soil types with different humidity.

4 class - in contains water surfaces and can be characterized with monotone decrease of the reflected ability of blue-green to red area of the spectrum, because wavelengths with longer length are absorbed stronger than the water.

IOP Conf. Series: Materials Science and Engineering **614** (2019) 012014 doi:10.1088/1757-899X/614/1/012014

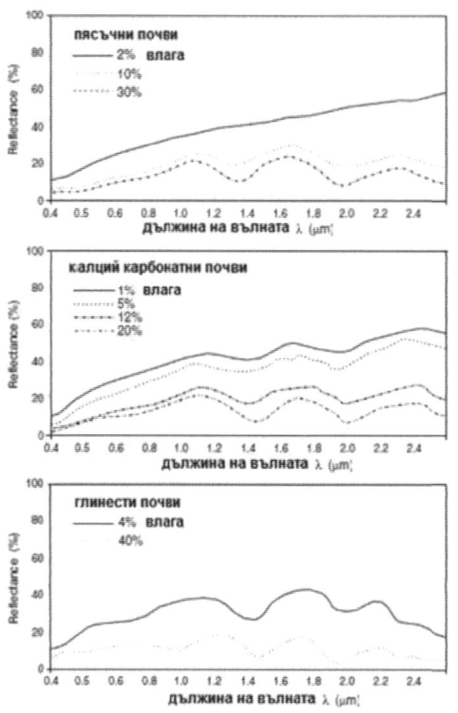

Figure 5. Spectral characteristics of soil types.

It must be noticed that data from hyperspectral ecological monitoring provide valuable and useful information about creation of library from spectral signatures for different natural and anthropogenic objects.

4. Systems for hyperspectral monitoring

The most reliable, accurate, operative and economic effective results at remote sense are produced when applying the so-called multiplier method. This is important for the hyperspectral ecological monitoring. In this aspect multiplier method suggests the combination usage of complex data, received from cosmic, aerial and on the ground hyperspectral sensors. For instance, on figure 6 can be seen organizational structure and basis components of the suggested system for hyperspectral monitoring [1].

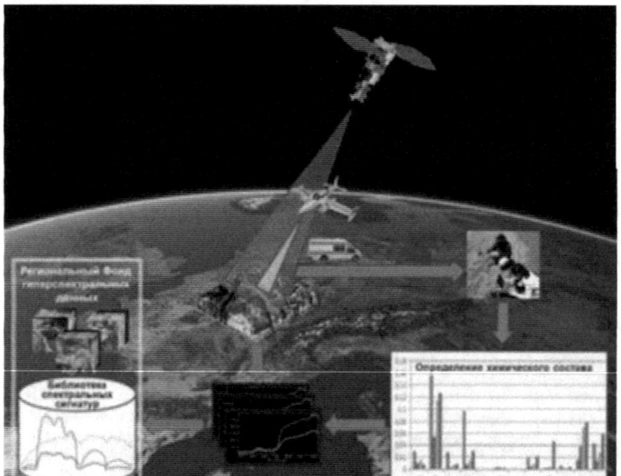

Figure 6. Structure of the system.

This multilevel system is based on using three levels, namely - cosmic, aerial and on ground. Its cosmic segment for operational hyperspectral monitoring is realized with spacecraft "Resurs-P". It is equipped with three types of shooting equipment, including hyperspectral [1]. The received different by types, but synchronously shot images of the same territory give a chance for improving the quality of delivered data and increasing their informational value. The aerial segment of this system, by its nature is kind plane- laboratory and can work in two mode. The first of them is so-called synchronous satellite image designed to calibrate hyperspectral data, delivered from the spacecraft segment. The second mode provide the receiving of operational hyperspectral information with high spatial resolution.

Due to technical respect, the aerial segment can be based on multitask plane or unnamed aerial craft. The last, the third on ground segment of the system include mobile and stational physics-chemical laboratories for getting soil samples, waters and others natural features to held their chemical analysis, as well as defining of the compliance between chemical composition of different substances and their spectral reflective characteristics, registrated onto the images. Data of every segment is necessary to create library of spectral signatures of different natural and anthropogenic objects.

The systems for operational hyperspectral monitoring provide non alternative data for receiving different final products as single thematic maps or geographical informational systems (GIS), which contain rich information for physics-chemical and ecological condition of different natural objects. These for example are data for: water quality at different water bodies, snow condition and future forecasts, vegetation surface condition, classification, soil ingredients, forest fires and area forecast for its distribution, existence of artificial and natural chemical and biological emission at the atmosphere and in water bodies, as well as forecast for consequences, presence of leaks from hydrocarbons, ammonia and other aggressive chemical products, sizes of areas covered with sludges from hydrocarbons and chemical products, ecological monitoring of industrial units from gases, petrochemical and chemical industries, rocks ingredients and areas of on ground geological researches, presence of archaeological objects located on the surface, presence of coastal borders and ecosystems and ext. Application of hyperspectral ecological monitoring are so much that is impossible to count. Below are given two visual examples - on figure 7 is shown ecological control of industry unit (town ot Omsk), on figure 8 - control of fires (Volga river).

Figure 7. Ecological control.

Figure 8. Fire control.

Not to forget that remote sense is object oriented, i.e. the solution of every applied and scientific tasks must be conformed with concrete feature of the studied objects.

5. Conclusion

To sum up, can be said that hyperspectral ecological monitoring must be considered as one innovative, convenient, high informative and economical effective system for delivering unique multilevel data for the real physics-chemical condition of the studied parts from the earth surface and located above it many natural and anthropogenically objects. It must be noticed that searching of hyperspectral information at the world market exceeds offering with 30 %. At the same time in Bulgaria, unfortunately, market of hyperspectral data is missing. One of the leading experts of the American society of photogrammetry and remote sensing Charles Mondedlo and George Hepner [6] give high rating form the current condition of the hyperspectral market in USA and forecast its stable dynamical development in the near future. Without doubt in world scale and in long-term perspective the meaning of hyperspectral monitoring will increase and operational data of real physics-chemical and ecological condition of our one and only home - the planet Earth.

References:

[1] Anshakov G P, Egorov A S, Rushkim A S and Schermunt V K 2013 Multifunctional Operational System Operations in Geomorphic Surveillance, *Samarskgo Gosdarstvennogo Aerokosmicheskogo universitet* **4** (42).

[2] Zdravcheva N D, 2016 *Proc. Int. Conf. EKO Varna (Varna)* vol. 23 (TU Varna) pp 46-52.

[3] Zdravcheva N D 2017 Spectral Transformations of Multi-satellite Satellite Images, *International Jubilee Scientific Conference 75th UACEG.*

[4] Zdravcheva N D Application of the Vegetation Indexes 2017 *25th International Symposium "Modern Technologies, Education and Professional Practice in Geodesy and Related Areas".*

[5] Mondello C Estimation of the Future US Market for Hyper X Data and Products URL: http://www.grss-ieee.org/wp- content/uploads/2009/10/Hyper-X-2010-v.1- 6.pdf.

[6] Shaw G Spectral Imaging for Remote Sensing 2003 *Lincoln Laboratory Journal* **14** p 3.

IOP Conf. Series: Materials Science and Engineering **614** (2019) 012015 doi:10.1088/1757-899X/614/1/012015

Flue Gas Desulfurization by Using Calcined Dolomite

S Baslayici[1,2], M Bugdayci[3], A Turan[3], L Oncel[4] and E Acma[1]
[1]Istanbul Technical University, Faculty of Chemistry and Metallurgy, Metallurgical and Materials Eng. Dep., 34469, Istanbul
[2]Istanbul Medipol University, Vocational School, Construction Technology Dep., 34810, Istanbul, Turkey
[3]Yalova University, Chemical and Process Engineering Department, Yalova, 77100, Turkey
[4]Sinop University, Faculty of Engineering and Architecture, Department of Metallurgical and Materials Engineering, 57000, Sinop, Turkey

Abstract. SO_2 emissions has begun to cause serious problems in industrialized countries. Today, various methods are used for removal of SO_2 gas. SO_2 gas is an important raw material for production of sulfuric acid. However, SO_2 content of flue gas must be over 6% to produce sulfuric acid commercially. Flue gases, which don't have enough amount of SO_2 to produce sulfuric acid commercially are harmful to the environment and human health and they must be filtered. In this study, calcined dolomite was used to remove SO_2 in flue gas which doesn't have enough SO_2 to produce sulfuric acid but have high emission values. Pyrite was used as SO_2 source and SO_2 removal efficiency of dolomite was determined for different experimental parameters. SO_2 capturing system was thermodynamically modeled and SO_2 capturing efficiency was determined as 77.81% in the experiments.

1. Introduction

The increasing population of the world and industrialization leads to an increase in energy needs. Use of fuels with sulfur content has been increasing to meet the rising energy demand. Emission of air-pollutant gases has been rising for that reason [1].

Sulfur dioxide (SO_2) is known as the most important air-pollutant gas. Main reasons that causes arising of SO_2 are natural sources and human activities. Microbial activities, volcanic eruptions, combustion of sulfur-containing fossil fuels, molten sulfur ores and oil refineries are among the causes of SO2 formation. [2-6].

SO2 is highly soluble in water and SO2 in flue gas is used as raw material in sulfuric acid production. On the other hand, SO2 content of flue gas must be over 6% to produce sulfuric acid commercially. It is necessary to filter the flue gases containing SO2 which do not have enough SO2 amount for production of sulfuric acid commercially. It may cause serious irreversible damage if they are released directly into the nature. [7-11]. SO2 gas pollutes the nature and it also increases the risk for cardiovascular diseases and lung cancer [12-14].

Sulfur dioxide is a colorless, sour-scented, fireproof and stifling gas. Taste of SO2 can be detected when its concentration reaches to 785 µgr/m3 and SO2's smell can be detected when its concentration reaches to 1305 µgr/m3 in the atmosphere. Sulfur dioxide is present in volcanic gases and its level is high in industrialized regions. SO2 is formed with combustion of sulfur-containing coals. This gas is condensed as a colorless liquid when cooled to -10 °C and it is solidified in white crystalline form when cooled to -73 oC [1, 15].

Researchers have been working on developing cost-effective systems for decreasing levels of air pollutant gases like sulfur dioxide (SO_2), nitrogen oxide (NO_X) and carbon monoxide (CO) [16-18]. In industry, gas-liquid reactions and gas-solid reactions are used for desulfurizing. Gas-liquid reactions are obtained by using activated carbon and liquid alkali materials (lime slurry and ammonia) [1, 17]. In other situations, catalysts (generally vanadium pentoxide) are used to turn SO_2 into sulfuric acid but these processes produce SO_2 gas as by product [19].

Main inorganic sorbent used in this process is calcite but in order to utilize large dolomite reserves in Turkey, dolomite is used as inorganic sorbent in this study. Results obtained from experimental studies were compared to the previous literature, where calcite and active carbon are used.

2. Experimental

In the experimental studies, pyrite ore from Kure region of Turkey was used as SO_2 source. Chemical composition of pyrite ore is given in Table 1 and XRD analysis result of pyrite ore is shown in Figure 1.

Table 1. Chemical composition of pyrite

Sample	Fe,%	S%	Cu%	Mg%	Ca%	SiO_2%	K%
Pyrite	38,18	41,67	0,58	0,38	0,77	9,33	0,02

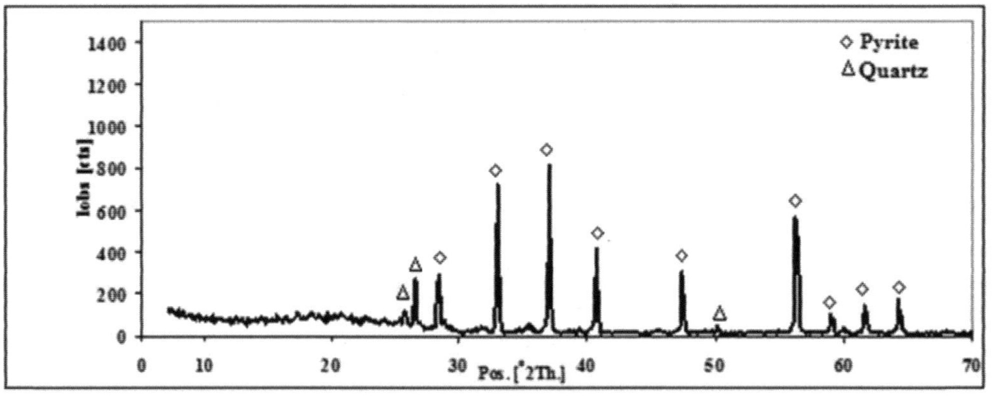

Figure 1. XRD analysis of Pyrite ore.

Mineralogical examinations show that pyrite (FeS_2), quartz (SiO_2) and hydronium jarosite ($(H_3O)Fe_3(SO_4)_2(OH)_6$) minerals were present in the pyrite concentrate. Main mineral of pyrite concentrate was pyrite and there were trace amount of chalcopyrite, covelline, marcasite, bornite, sphalerite, digenite and various silica minerals. Acoording to particle size measurements, mean particle size of pyrite was 41 μm.

Calcined dolomite used in experimental studies was Kutahya Yaylababa dolomite and it was procured from Kumas Magnezit Sanayi Corporation. Chemical analysis of Yaylababa dolomite is shown in Table 2. Calcined dolomite's mean grain size was 42 μm.

Table 2. Chemical analysis of dolomite.

Sample	Loss on Ignition	CaO%	MgO%	FeO%	Al_2O_3%	SiO_2%	Na_2O%
Dolomite	9.52	57.59	31.00	0.12	Trace	1.07	0.0086

Thermo Scientific XSeries 2 ICP-MS was used for chemical analyses of the samples. Bruker D8 Advance was employed for XRD analyses and mineralogical examinations. Roasting operation were carried out in a tube furnace working with 220V electricity. Samples were grinded with ring mill and dehumidified in Binder drying-oven. Samples particle size were measured with Malvern Mastersizer 2000.

In the roasting experiments, ceramic boat was weighed initially and it was weighed again after filling with pyrite. Pyrite was roasted in furnace and weight loss caused by roasting was measured by weighing. Roasting process was carried out with different times and optimum roasting time was determined. Various air flow rates were used in SO_2 capturing experiments. SO_2 emissions were measured after roasting process for every air flow rate used. SO_2 emission measurements after roasting process were repeated for different flow rates with using dolomite for capturing SO_2 gas. Then, SO_2 capturing results were compared.

Titrimetric method was used in SO_2 analysis. In this method, gases released from the flue were kept in the hydrogen peroxide solution which was in the washing bottle and it was converted to sulfuric acid. Concentration of SO_2 was determined by titration with a base with a known normality.

100 ml. of a 1% H_2O_2 solution was filled in each of the two gas washing bottles, which were coupled in series. Gas exited from the flue was transferred to first gas washing bottle. SO_X was oxidized to acid by H_2O_2 in the washing bottle. SO_X gas, which didn't react in the first washing bottle moved to second washing bottle and reacted with H_2O_2. After the process was completed, solutions in two washing bottles were titrated with alkali (1N Na_2CO_3) and equivalent SO_2 quantity in the flue gas was determined. Reactions took place in this process are given in Equation 1 and Equation 2.

$$SO_2 + H_2O_2 \rightarrow SO_3 + H_2O \quad \textbf{(1)}$$

$$SO_3 + H_2O \rightarrow H_2SO_4 \quad \textbf{(2)}$$

Total reaction is shown in equation 3.

$$SO_2 + H_2O_2 \rightarrow H_2SO_4 \quad \textbf{(3)}$$

Acid's titration reaction is given in Equation 4.

$$H_2SO_4 + Na_2CO_3 \rightarrow Na_2SO_4 + H_2CO_3 \quad \textbf{(4)}$$

SO_2 concentration in the gas mass is calculated as shown in Equation 5. Here, N is normality of the base and S is base consumption (ml).

$$SO_2\,(mg/l) = (32.\ N_{Na2CO3}.\ S_{Na2CO3})\ /\ \text{volume of air (l)} \quad \textbf{(5)}$$

SO_2 capturing experiments were started after determining the roasting conditions in the first experimental set. Calcined dolomite was used to capture SO_2 in this experimental set. Different air flow rates were used in SO_2 capturing experiments. Initially, SO_2 formation amounts were determined for every air flow rate used. Then, dolomite was used to capture SO_2, and capturing efficiencies were compared. The amount of dolomite was determined by weighing the empty boat and then weighing the boat filled with dolomite. The resulting products were weighed and compared with roasting experiments' results.

3.Results and Discussion

In the experimental studies, quantity of theoretical air requirement for roasting of pyrite was calculated initially. Studies were carried out with two times, four times, six times and eight times of the theoretical air quantity. According to Equation 6; in order to obtain 0.0651 mole S, 0.31 mole of air is required. Table 3 represents roasting experiments' conditions.

$$S + O_2 \rightarrow SO_2 \quad \textbf{(6)}$$

IOP Conf. Series: Materials Science and Engineering **614** (2019) 012015 doi:10.1088/1757-899X/614/1/012015

Table 3. Roasting experiments' conditions.

Exp. No	Duration (min.)	Flow Rate (l./min.)	Weight Before Roasting (g.)	Weight After Roasting (g.)	Difference (g.)	SO_2 Amount (mg./l.)
1			5.79	3.93	1.86	32.15
2	15		5.78	3.93	1.85	32.08
3			5.30	3.60	1.70	32.10
Avg.15			**5.62**	**3.82**	**1.80**	**32.11**
4			5.34	3.44	1.89	35.52
5	30		5.25	3.39	1.86	35.47
6			5.33	3.43	1.90	35.68
Avg. 30		1	**5.30**	**3.42**	**1.88**	**35.55**
7			5.17	3.34	1.83	35.44
8	45		5.32	3.43	1.88	35.36
9			5.31	3.42	1.88	35.47
Avg. 45			**5.26**	**3.39**	**1.86**	**35.42**
10			5.34	3.44	1.89	35.53
11	60		5.29	3.41	1.87	35.45
12			5.17	3.33	1.83	35.50
Avg. 60			**5.26**	**3.39**	**1.86**	**35.49**

Experiments were repeated three times for all roasting conditions and average values were calculated. Roasting experiments showed that 1 l./min. air flow rate and 2 times of the theoretical air quantity was not enough to complete the roasting process. Roasting efficiency was increased to 100% with four times of the theoretical air quantity used and it was determined that using more air in the process was not necessary.

Figure 2 shows effect of roasting time on SO_2% formation. It is seen in the figure that amount of SO_2% was increased exponentially when roasting time was increased from 15 minutes to 30 minutes. SO_2% amount remained nearly constant when roasting time exceeded 30 minutes. Table 3 and Figure 2 shows that optimum roasting conditions of pyrite is 30 minutes of roasting time and 1 l./min. of air flow rate.

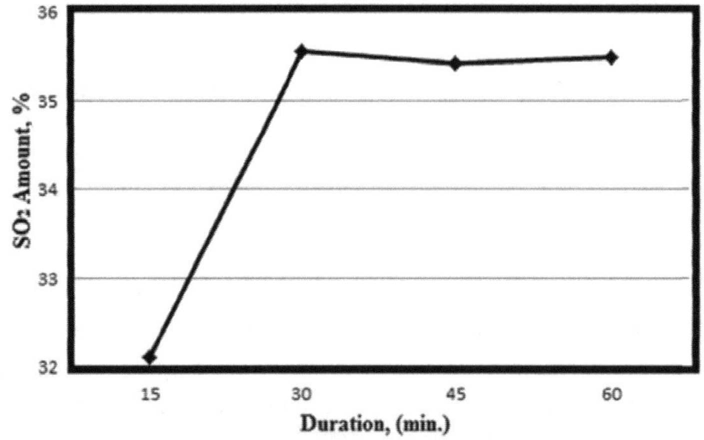

Figure 2. Change of SO_2% amount versus time.

After roasting experiments, SO_2 capturing system was thermodynamically modeled via FactSage 6.2 database. Figure 3 illustrates standard free energy of dolomite and pyrite concentrates changing with temperature. Figure 3 shows that dolomite will be sulphated and pyrite will be oxidized at 600 °C.

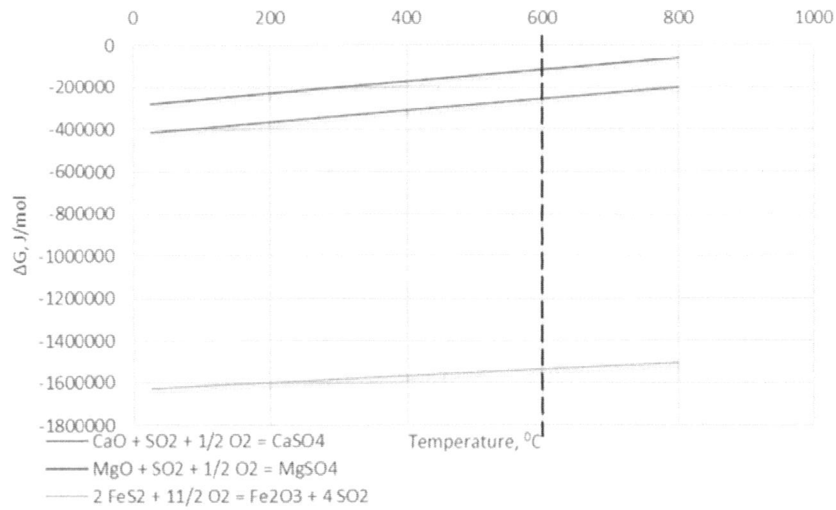

Figure 3. Standard free energy of dolomite-pyrite system changing with temperature.

Two boats were used in SO_2 capturing experiments. Pyrite concentrate was charged in the first boat and dolomite was charged in the second boat. Table 4 presents experimental parameters and results of SO_2 capturing experiments. SO_2 capturing amount versus time curve, which was created by using data in Table 4 is shared in Figure 4.

In SO_2 capturing experiments, better results were obtained at low air flow rates. Contact time between solid and gas is increased at low air flow rates. Contact surface area, which is needed for occurring of reactions remain same but contact time is increased with using low air flow rates. Therefore, flue gas desulfurization efficiency increases at low air flow rates. The optimum SO_2 capturing result was obtained as 77.76% when 0.4 l. /min. air flow rate was used.

Effect of temperature on SO_2 capturing efficiency was investigated by using different temperatures with constant air flow rate. Results showed that temperature didn't have a significant effect on capturing efficiency. On the other hand, high temperatures were used in experimental studies and it doesn't provide information about low temperatures like room temperature. Figure 4 shows that SO_2 capturing efficiency increased insignificantly from 77.76% to 77.81% with increasing temperature. According to these data, flue gas desulfurization efficiencies of the companies using sulphurous fuel will show insignificant changes due to the narrow temperature range of the flue gases.

When all results are taken into consideration, filtering with dolomite is suitable for plants with average emission values lower than 1285µg per hour, 485 µg per day and 214 µg per year. Mangun et al. obtained 82% SO_2 capturing efficiency by using active carbon but large dolomite reserves in Turkey make dolomite a good alternative to active carbon. In this study, SO_2 which is captured by dolomite can be converted into sulfuric acid [20]. Mathieu et al.'s study showed that SO_2 can be captured by using calcite but regeneration of captured SO_2 is not possible in this method. Therefore, using dolomite instead of calcite is advantageous [21].

IOP Conf. Series: Materials Science and Engineering **614** (2019) 012015 doi:10.1088/1757-899X/614/1/012015

Table 4. Conditions and results of SO_2 capturing experiments.

Exp. No	Temperature, °C	Flow Rate, lit./min	Weight Before Roasting, g.	Weight After Roasting, g.	SO_2 Amount, mg./lt.	SO_2 Capture, %
1			1.07	0.68	25.60	-
2			1.07	0.69	26.67	-
3	600	1.0	1.09	0.70	25.60	-
Avg.			**1.07**	**0.69**	**25.95**	-
4			1.00	0.64	7.46	71.25
5			1.21	0.77	7.46	71.25
6	600	1.0	0.99	0.64	6.93	73.29
Avg.			**1.06**	**0.68**	**7.28**	**71.93**
7			1.06	0.68	33.33	-
8			1.15	0.73	33.33	-
9	600	0.8	1.05	0.68	32.67	-
Avg.			**1.08**	**0.69**	**33.11**	-
10			1.07	0.69	8.67	73.81
11			1.09	0.70	8.67	73.81
12	600	0.8	1.02	0.66	9.33	71.82
Avg.			**1.06**	**0.68**	**8.89**	**73.14**
13			1.16	0.75	44.44	-
14			1.04	0.66	44.89	-
15	600	0.6	1.10	0.71	44.89	-
Avg.			**1.10**	**0.70**	**44.59**	-
16			1.10	0.71	10.67	76.15
17			1.00	0.64	11.11	75.16
18	600	0.6	1.02	0.65	10.22	77.15
Avg.			**1.04**	**0.66**	**10.66**	**76.15**
19			1.06	0.68	64.00	-
20			1.12	0.71	64.00	-
21	600	0.4	1.01	0.65	64.00	-
Avg.			**1.06**	**0.68**	**64.00**	-
22			1.02	0.67	14.67	77.07
23			1.11	0.70	14.67	77.07
24	600	0.4	0.99	0.63	13.34	79.15
Avg.			**1.04**	**0.66**	**14.22**	**77.76**
25			1.05	0.67	64.00	-
26			1.05	0.68	61.34	-
27	700	0.4	1.05	0.68	64.00	-
Avg.			**1.05**	**0.67**	**63.11**	-
28			1.09	0.70	14.67	76.75
29			1.11	0.68	13.67	78.34
30	700	0.4	1.03	0.66	13.67	78.34
Avg.			**1.07**	**0.68**	**14.00**	**77.81**

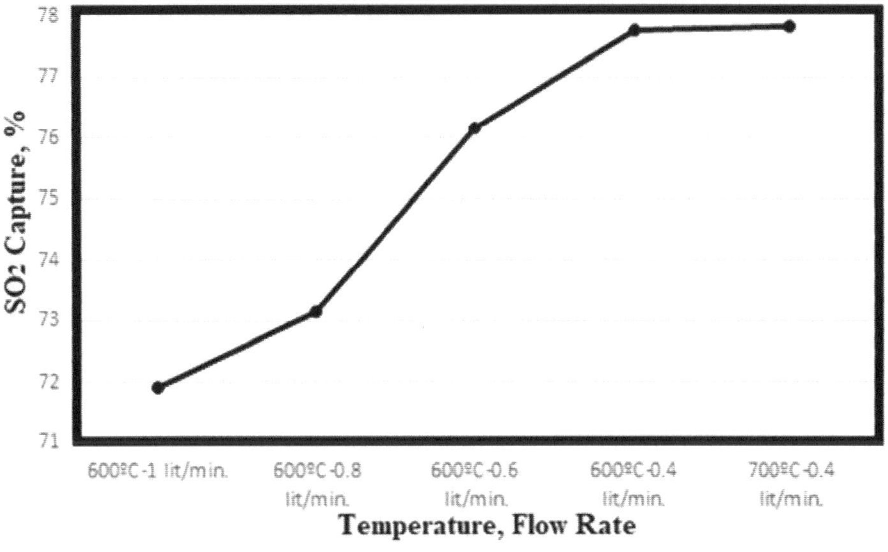

Figure 4. Effect of temperature and air flow rate on SO_2 capturing amount.

4.Conclusion

The world's energy requirement has been increasing because of the development in technology and industries. High SO_2 emissions caused by use of fuels with sulfur content must be reduced. SO_2 emission values are high in Turkey and it continues to increase despite the downward trend in developed countries. SO_2 emission values of Turkey should be reduced in the short term and the long term.

Because of the harmful effects of SO_2 gas on nature and human health, flue gases which don't have enough amount of SO_2 to produce sulfuric acid commercially but above the limits set by the authorities must be filtered.

In experimental studies, mineralogical analysis of Yaylababa dolomite was performed initially. Then, thermodynamic calculations were carried out and it was seen that dolomite can capture SO_2 gas. Experiments showed how theoretical calculations were working in practice.

Roasting conditions of pyrite concentrate were determined by experiments. The optimum air quantity required was calculated theoretically and determined in practice. It was seen that using four times of the theoretical air quantity was enough and there was no need to use more air. Optimum conditions vary according to the content of the concentrate and it can be specified as two to four times the theoretical amount of air.

In the experiments, SO_2 capturing efficiency of dolomite was reached to 77.76% at optimum conditions. However, it is predicted that efficiency value may decrease in small amounts at high air flow rates.

5. References

[1] Baslayici S 2014, Flue Gas Desulphurization by Using Calcined Dolomite, MSc Thesis, Istanbul Technical University, Turkey.
[2] Li N, Wang Z and Zhang Z 2018, *Environ Sci Pollut Res* 25 (35) 35099-35108.
[3] Fowler D and Unsworth M H 1977, *Quart. J. R. Met. SOC* 105 767-783.
[4] Hu K, Bai G, Li W, Yan H, Hu, Li L and Zhang H 2014, *Plant Growth Regul* 75 (1) 271-280.
[5] Li P, Wang X, Allinson G, Li X, Stagnitti F, Murray F and Xiong X 2011, *Environ Sci Pollut Res* 18 (7) 1090-1097.
[6] Majernik O and Mansfield T A 1970, *Nature* 227 377-378.

IOP Conf. Series: Materials Science and Engineering **614** (2019) 012015 doi:10.1088/1757-899X/614/1/012015

[7] Blakemore F B, Davies C and Isaac J G 1999, *Appl Energy* 62 (4) 283-295.
[8] Fenger J 2009, *Atmos Environ* 43 (1) 13-22.
[9] Harnisch J, Borchers R and Fabian P 1995, *Environ Sci Pollut R* 2 (4) 229-232.
[10] Shepherd M A, Haynatzki G, Rautiainen R and Achutan C 2015, *J Air Waste Manag Assoc* 65 (10) 1239-1246.
[11] Taieb D and Brahim A B 2013, *C R Chim* 16 (1) 39-50.
[12] Zhang D, Liu J, Jiang C, Li P and Sun Y 2017, *Sens Actuators B Chem* 245 560-567.
[13] Chaudhary V, Kaur A 2015, *Polym. Int.* 64 (10) 1475-1481.
[14] Liu X Y, Zhang J M, Xu K W and Ji V 2014, *Appl Surf Sci* 313 405-410.
[15] Azimi M, Feng F and Yang Y 2018, *Sustainability* 10 (2) 367.
[16] Miller M J 1986, *Environ Prog Sustain Energy* 5 (3) 171-177.
[17] Wark K, Warner C F, Davis W T 1998 *Air Pollution: Its Origin and Control* 3rd Edition, (California: Addison-Wesley), p. 573.
[18] Nersesian R L 2007, *Energy for the 21st Century: A Comprehensive Guide to Conventional and Alternative Sources*, (New York: M. E. Sharpe) New York, p. 456.
[19] Bauer G, Güther V, Hess H, Otto A, Roidl Roller O H, Sattelberger S, Köther Becker S, and Beyer T 2017 *Vanadium and Vanadium Compounds* (In Ullmann's Encyclopedia of Industrial Chemistry), (Online: Wiley VCH), p. 22.
[20] Mangun C L, Debaar J A and Economy J 2001, *Carbon* 39 1689–1696.
[21] Mathieu Y, Tzanis L, Soulard M, Patarin J Vierling M and Molière M 2013, *Fuel Processing Technology* 114 81–100.

IOP Conf. Series: Materials Science and Engineering **614** (2019) 012016 doi:10.1088/1757-899X/614/1/012016

Impact of road transport on the pollution of ambient air with particulate matter

Uzuntonev T*, Penchev B

Department "Automotive engineering", Faculty of Mechanical Engineering and Technologies, Technical University of Varna, 1"Studentska" Str, Varna, 9010 Bulgaria

*E-mail: uzuntonev_trifon@abv.bg

Abstract. In many cases some media analyses of the influence of vehicles with internal combustion engines on the air pollution show that particulate matter with a size of PM10 and PM2.5 are over exaggerated. This problem in the publications is analysed from technical, rather than from emotional point of view. The tests are made for particulate matter with a size of PM10 and PM2.5 on a main road with intensive traffic in an average European city. The experiment in conducted in the city of Varna, Bulgaria, with a population of 380,000 citizen. The influence of different factors, such as humidity, distance from road and green areas is considered under the same road traffic. It is proven that exactly the above mentioned indirect factors are crucial for the level of pollution. There are different examples of household activities which particulate matter emissions are much higher that the permissible norm.

1. Introduction

The problem of ambient air quality and the impact of road transport on this process is extremely important, particularly if it concerns typical urban environment. Similar topic is contained in much of the media commentaries on this problem. Usually, in most cases diesel vehicles are indicated as key pollutants of ambient air. In many cases, no distinction is made between eco-standards, restrictive levels of toxic components in exhaust gases for different types of engines, and dates of entry into force of the regulations. Such interpretation of the issue is emotional, even politically tolerated, but is rarely based on certain technical facts and does not take into account the achievements of automobile manufacturers in their efforts to reduce the levels of exhaust gas toxicity emitted into the atmosphere [1,2].

The report analyses the impact of road traffic in an average European city on pollution of ambient air with particulate matter (PM). The research was conducted in the city of Varna, Bulgaria (with a population of 380,000 inhabitants) where traffic control and intensity largely approximated those specified in WLTP standard, in particular its urban traffic cycle part. The level of PM pollution was measured, taking into account a number of additional factors, such as location of measurement in the pedestrian zone, infrastructure quality, air humidity, landscaping of the surrounding areas and other additional factors which form the real environmental picture that is often quite different from the laboratory conditions of environmental standards.

2. Explanation

The adverse effects of particulate matter on human health have been investigated for a long time. The interest of researchers is redirected from PM to PM10 and PM2.5 or to even finer one. The source of this matter includes mainly of primary aerosols generated from anthropogenic activities such as transport, industry, construction and secondary aerosols. In large cities, traffic on asphalt roads significantly contributes to this. Particulate matter is one of the main air pollutants [3]. Its harmful health effect depends largely on the size and chemical composition of suspended particulate matter, other chemical compounds absorbed on its surface, including mutagens, DNA modulators, etc., and on the area of respiratory system in which they are deposited. The main sources of dust are industry, transport and energy. Particulate matter enters the body mainly through the respiratory system (figure 1), where the larger PM is retained in the upper respiratory tract, while the finer PM (below 10 mm - PM10) reaches the lower respiratory system sections, causing damage to the lung tissues. Children, adults and people with chronic lung disease, influenza or asthma are particularly sensitive to high PM10 levels.

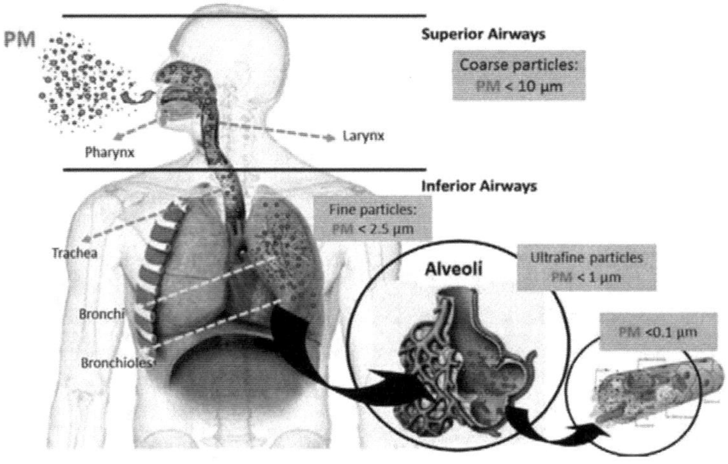

Figure 1. Absorption of solid particulates by the human body [4].

Short-term exposure to 500 mg / m^3 of dust and Sulphur dioxide increases the overall mortality rate in the population, while the exposure to half-lower concentrations increases in morbidity and impairment of pulmonary function. Prolonged exposure to Sulphur dioxide and dust leads to increase in non-specific pulmonary diseases, mainly respiratory infections of the upper respiratory tract and bronchitis at significantly lower concentrations (30-150 mg / m^3), which is particularly pronounced in children. Patients with chronic bronchial asthma and cardiovascular diseases are most vulnerable to the combined effects of dust and Sulphur dioxide.

The legislation of the Republic of Bulgaria in this direction is presented by Ordinance 12 of 15 July 2010 (promulgated in State Gazette No 58 of 30 July 2010) on setting concentration limit values (CLV) for particulate matter. The introduced CLVs aim to prevent PM harmful effects on human health and environment. The following standards for particulate matter are defined: PM10

- Daily average rate - 50 mcg / m^3 (this rate should not to be exceeded more than 35 times a year), annual average rate - 40 mcg / m^3

Such limitations in the form of absolute values do not give an idea of the differentiated origin of measured PM and, in particular, this associated with road transport. To clarify this influence, a number of measurements were made on the territory of the city of Varna, Bulgaria. At the presentation of the results, a section of a ring road with high intensity of mass movement of vehicles was selected. The section is characterized by different infrastructure and landscaping in both directions of traffic. This makes it possible to take into account additional factors influencing the process of pollution of ambient air with PM. Figure 2 shows the location of the measurements. Pavement and green areas next to it are located in the southern part of the lane. The northern part lacks pavement; instead, there are concreted areas with high level of dust. After a west-east movement, the existing slope that causes engine load should be overcome. In the opposite direction, the conditions imply lower load on the drive unit.

Figure 2. Location in conducting PM2,5 and PM10 emission monitoring - Varna Ring Road (East-West).

There are no industrial plants and other suspected pollutants in the vicinity of the measurements. The annual measurement period excludes the presence of a superimposed background from combustion of environmentally unfriendly solid fuels for domestic heating. It is therefore expected that the level of pollution with PM thus measured is entirely due to road traffic. It should be stressed that air pollution with PM caused by road traffic is complex and reflects the exhaust gas pollution, turbulence of particulate matter from vehicle movement, particles emitted by the braking system, products from the wear of road surface, marking, and tires and vehicle transmission.

In this respect, it is difficult to make a differentiated assessment of the impact of PM pollution from various vehicles propelled by gasoline, diesel or internal combustion engines or by vehicles with purely electric drive. The extremely important factor - geographical location of the settlement - remains constant. In this case, it is about a settlement on the Black Sea coast with year-round air flows. This largely explains the fact that city of Varna rarely appears in the environmental information on the increased level of PM in the air. Settlements located in basins with unfavorable atmospheric pressure are much more threatened by increased particulate matter concentrations. Last but not least, the weather conditions and, above all, the presence of rainfall or high humidity, must be taken into account. It is noteworthy that under these conditions the level of PM in the air falls sharply.

The analysis of the variety of factors influencing the total PM quantity in the air emitted at a given moment excludes the objectivity of extreme assessments that certain type of pollutants are dominant. After considering all these peculiarities, measurements are made in two consecutive weekdays at the same time, but at different air humidity. Table 1 and Table 2 show the results obtained in a sunny day with no rainfall at a wind speed of up to 0.1 m / s. To better perceive the environmental situation, the pedestrian zone is divided in the form of a measuring network with a sample rate of 1m. (figure 3).

Figure 3. Network of measured points.

Measurements of PM2.5 and PM10 are made at these points of the raster network. Figure 3 shows the silhouettes of a woman and a man with an average statistical height. In all cases, the PM level is far below the maximum average daily rate in the inhalation area. If, however, the measurement is made in the immediate proximity of the vehicle wheels where air flow turbulence is dominated, higher air pollution values, even values exceeding the maximum rates may be reported. Such measurements of practically the same environmental situation may lead to subjective interpretation and even diametrically opposed conclusions. Table 3 and Table 4 show the same road situation, but on a weekday with low rainfall and high air humidity. In this case, the sharp reduction in PM content is obviously due to the more favorable weather conditions.

Table 1. Values reported at a dry road - left side (vegetation).

Parameter	**PM2,5** /PM10	**PM2,5** /PM10	**PM2,5** /PM10	**PM2,5** /PM10
Moving away	x	x+1 meter	x+2 meters	x+3 meters
3 m height	**13**/10	**9**/8	**6**/6	**3**/2
2 m height	**24**/19	**12**/10	**17**/8	**10**/4
1 m height	**38**/32	**23**/18	**17**/14	**9**/6
at road (x)	**73**/68	**29**/26	**17**/16	**11**/8

Table 2. Values reported at a dry road - right side (buildings).

Parameter	**PM2,5** /PM10	**PM2,5** /PM10	**PM2,5** /PM10	**PM2,5** /PM10
Moving away	x	x+1 meter	x+2 meters	x+3 meters
3 m height	**13**/10	**9**/8	**5**/6	**5**/4
2 m height	**21**/18	**17**/14	**15**/12	**11**/10
1 m height	**44**/37	**28**/24	**23**/21	**15**/14
at road (x)	**92**/80	**48**/42	**35**/28	**20**/16

Table 3. Values reported at a wet road - left side (vegetation).

Parameter	**PM2,5** /PM10	**PM2,5** /PM10	**PM2,5** /PM10	**PM2,5** /PM10
Moving away	x	x+1 meter	x+2 meters	x+3 meters
2 m height	**12**/10	**9**/8	**7**/6	**5**/4
1 m height	**19**/16	**9**/8	**13**/11	**10**/9
at road (x)	**29**/25	**19**/16	**12**/10	**8**/7

Table 4. Values reported at a wet road - right side (buildings).

Parameter	**PM2,5** /PM10	**PM2,5** /PM10	**PM2,5** /PM10	**PM2,5** /PM10
Moving away	x	x+1 meter	x+2 meters	x+3 meters
2 m height	**12**/10	**12**/10	**13**/11	**8**/7
1 m height	**32**/27	**10**/9	**14**/12	**15**/13
at road (x)	**33**/29	**18**/15	**14**/12	**8**/7

The dramatic presentation of the role of road transport in pollution of air with PM provoked the research team to measure the levels of this component in common everyday household activity. Figure 4 presents the results of the illustrated activities. All of them drastically exceed the maximum daily rates, on the one hand, but they have always been an accompanying element of the human lifestyle, which the people perform voluntarily, on the other hand.

Lighted cigarette PM10 up to 300 µg/m^3

Barbecue PM10 over 1,200 µg/m^3

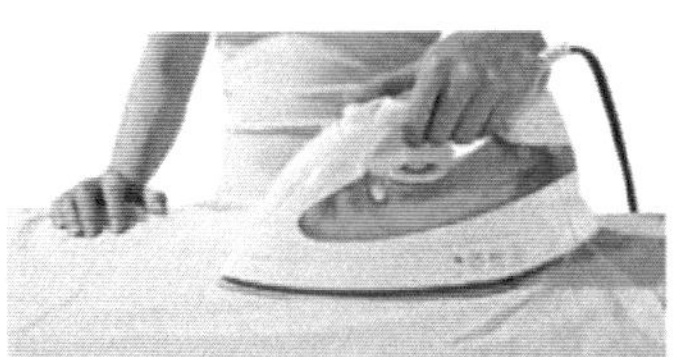

Clothes ironing PM10 up to 300 µg/m^3

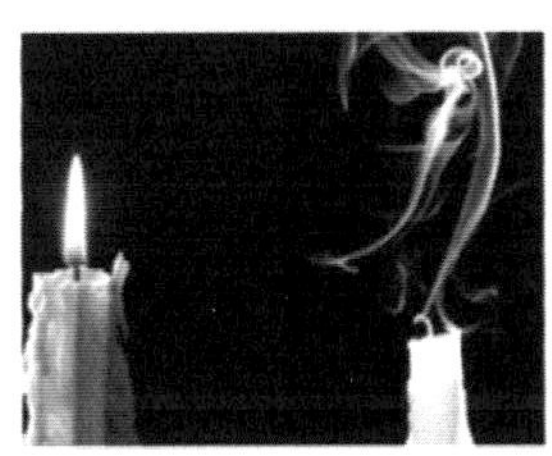

Lighted candle PM10 from 110 to 220 µg/m^3

Figure 4. Household activities.

3. Conclusion

1. Ten percent of the total particulate matter emitted during the operation of a EURO 6 compliant vehicle is attributed to the exhaust gases.

2. Additional factors influence the total PM in the operation of road transport, besides the exhaust gases. These are:

-air turbulence caused by the vehicle movement;

-infrastructure quality;
-air humidity (presence of rainfall);
-landscaping of the surrounding areas;
-Geographical location of the settlement.

3.A number of household activities cause PM to be released at levels that are significantly higher than those of a modern vehicle's exhaust gases.

References

[1] Thomas K 2017, Karlsruhe Der Dieselmotor: Nüchterne Bewertung der Emissionsthematik.
[2] Thomas K 2014 Karlsruhe Der Verbrennungsmotor 2050 – Antrittsvorlesung zum Festkoiioguium der IFKM *(*Institut für Kolbenmachinen*).*
[3] Dimitrov A, Sevastakiev V 2006 *Environmental Characteristics of Internal Combustion Engines and Vehicles* (Varna)
[4] Guarieiro L L and Guarieiro A L 2013 Vehicle Emissions: What Will Change with Use of Biofuel? *Biofuels - Economy, Environment and Sustainability* (ed. Zhen Fang) (IntechOpen) DOI: 10.5772/52513.

Institute of Physics
Dirac House, Temple Back
Bristol BS1 6BE UK

ISSN: 1757-8981
ISBN 978-1-5108-9821-9

9 781510 898219

CURRAN ASSOCIATES INC.
proceedings.com